Evolution and Human Culture

Value Inquiry Book Series

VOLUME 290

Cognitive Science

Edited by

Francesc Forn i Argimon

The titles published in this series are listed at *brill.com/cosc*

Evolution and Human Culture

Texts and Contexts

By

Gregory F. Tague

BRILL
RODOPI

LEIDEN | BOSTON

Cover image: Bisson Frères (French, active 1840–1864) Louis-Amédée Mante (French, 1826–1913). Photolithograph by Lemercier et Cie. [Gorilla], about 1853–1854, Salted paper print 27.6 × 19.5 cm (10 7/8 × 7 11/16 in.). The J. Paul Getty Museum, Los Angeles. Digital image courtesy of the Getty's Open Content Program.

Library of Congress Cataloging-in-Publication Data

Names: Tague, Gregory, author.
Title: Evolution and human culture : texts and contexts / by Gregory F. Tague.
Description: Boston : Brill-Rodopi, 2016. | Series: Value inquiry book series, ISSN 0929-8436 ; VOLUME 290. Cognitive Science | Includes bibliographical references and index.
Identifiers: LCCN 2016011383 (print) | LCCN 2016013370 (ebook) | ISBN 9789004305366 (pbk. : alk. paper) | ISBN 9789004319486 (E-book)
Subjects: LCSH: Ethics, Evolutionary. | Culture. | Cognition. | Aesthetics.
Classification: LCC BJ1311 .T34 2016 (print) | LCC BJ1311 (ebook) | DDC 155.7--dc23
LC record available at http://lccn.loc.gov/2016011383

Typeface for the Latin, Greek, and Cyrillic scripts: "Brill." See and download: brill.com/brill-typeface.

ISSN 0929-8436
ISBN 978-90-04-30536-6 (paperback)
ISBN 978-90-04-31948-6 (e-book)

This book is printed on acid-free paper and produced in a sustainable manner.

For Fredericka and Karolina

"With monkeys and apes around every corner, no rain forest culture has ever produced a religion that places humans outside of nature."
Frans de Waal, *The Age of Empathy*

Contents

Preface

Why do some people get heated up when they hear the word evolution? Why are some people incensed about any hint of human evolution? Why do some people accept the evolution of other animals but not of human beings? Why do some people accept some version of physical human evolution but resist any suggestion that behaviors such as morality and art, or that the human mind, are evolved? In 2005 we began seeing articles reporting the results of surveys concerning the so-called public acceptance of evolution. Jerry Coyne and J. Miller discuss these trends. For example, in America alone in 1985 forty percent of the population did not believe in the veracity of evolution, and that percentage of non-believers has increased to forty-five percent in recent years.

My book not only accepts the truth of evolution, but the truth that human behaviors, practices, beliefs, values, emotions, and reasoning are evolved adaptations, though expressed differently in various cultures. Taking the view of Andrew Shryock and Daniel Lord Smail, I argue for a full study of human prehistory utilizing various disciplines. Our mind is responsible for behavior and has many adaptations, says David Buss, especially in our interpersonal actions.

Evolution and Human Culture will be valuable to students and scholars of literature, the arts, and cultural studies, including moral philosophers, who would be interested in reading about key intellectual developments in their fields. Biologists and social scientists would benefit as well, since the book provides a window into how scientific research contributes to the arts and humanities. The book offers a comprehensive entry into evolutionary cultural studies. The take-home point is that culture does not transcend nature; culture is human nature with moral sensations at bottom.

Most professors and students are still spinning abstractions with postmodernist theories, but they need to consider a more empirical approach. For instance, how could moral philosophers ignore biology? Most do, and I suppose in some cases that is justifiable. How could scholars writing definitive papers on beauty ignore neuroscience? Many do, and I guess in some cases that is justifiable. But these are omissions of ignorance. As another instance, theorists think that culture is something separate, foisted upon us from elsewhere, indicating that we are born with nearly empty minds that need to be filled. The notion of social constructivism is for the most part incorrect.

Evolutionary psychologists do not ignore the impact of the complex function of environment, but to a greater degree we need to see that culture is an evolved behavior so that some of our cultural imperatives are inborn. While in our advanced state of civilization cultural practices have become runaway, at bottom their existence answered adaptive needs.

Broadly speaking there are several aims of this book:

- To survey and concisely review some literature in various disciplines of evolutionary studies in order to collect it into one volume for a wider audience.
- To demonstrate, especially for those working in the arts and humanities, how an evolutionary approach could dramatically enlighten and invigorate their studies.
- To suggest that much of our cultural production stems from what in early hominins was a caring tendency – both the care to share and a self-care to challenge others.
- To argue that cognition and feelings gave rise to cognitive emotions crucial for adaptations regarding individual well-being and strategizing in the context of group norms.
- To argue that cultural artifacts as social expressions are physical examples of individual and shared moral sentiments.
- To conclude that while culture is a controlling mechanism to set boundaries, at the same time it invites innovation to fire change.

The model of mind followed is that of Steven Mithen and cognitive fluidity. Mithen builds from Howard Gardner's notion of intelligences and Leda Cosmides and John Tooby's idea of modular flexibility. What many call our social brain is really a reflective mind adapted to sentiments. We think creatively through our emotions. Cognitive skills in monkeys and great apes evolved in terms of social pressures. With us, emotional cognition magnified to the point that our culture is highly protean. While nascent in apes, our mentality includes a flourishing ability to attribute mental states to others and to hypothesize social situations. So our culture and aesthetics come in line with what I term moral.

Our brains are clearly much less domain specific than those in monkeys and to some extent great apes, precisely because we are concerned about what others will think of our actions. This book does not, technically, deal with how the mind works. Rather, with the evolution of human culture there is less concern about a model and more about referencing chronology, artifacts, mental adaptations, and continuities with nonhuman primates. While culture consists of laws, codes are meant to be broken. Our ability to mold a letter of the law and to innovate over norms is the hallmark of our cultural behaviors.

The book is broken down into four main sections, asserting that essentially everything human is cultural, and everything cultural is cognitive or emotional. The sequence of the chapters mimics the argument about how artistic representations and cognitive culture challenge and yet enforce prehistoric and adaptive moral norms. The trajectory of the book builds to art and aesthetics, the culmination of what it means for us to be human. And yet, chapter two rightly focuses on a key tenet of the book and of humanity, our moral sentiments.

My more comprehensive and detailed study *Making Mind: Moral Sense and Consciousness*, deals with the origin and evolution of individual consciousness as narration. In that book I address more topics more deeply, such as moral sensations and emotions, consciousness, free will, and individual/group selection. That book and this one complement each other and could be read together, since here I focus more on what is called artistic culture, which includes cultural practices and values as well as the arts. The term culture is too broad, and the word art is too narrow, so I combine the two. In the creation of and participation in artistic culture, I will refer to art behavior.

Acknowledgements

Parts of this book appeared originally in the form of reviews in *ASEBL Journal* or *Consciousness, Literature and the Arts*. Permission to reproduce material in modified and substantially revised form has been granted by the editors of those journals.

Cover: Digital image courtesy of the Getty's Open Content Program.
Gorilla, Bisson Frères, circa 1853.

To St. Francis College for granting me a sabbatical in which to substantially finish this project.

To my editors, Professor Francesc Forn i Argimon and Eric van Broekhuizen for their guidance and support.

To the following scholars for valuable comments and suggestions: Kathryn B. Francis, CogNovo Fellow, Plymouth University; Christopher X.J. Jensen, Associate Professor of Ecology and Evolution, Pratt Institute; Daniel Meyer-Dinkgräfe, Professor, Lincoln School of Fine and Performing Arts; Steven Ross, Professor of Philosophy, Hunter College and the CUNY Graduate Center.

To Kasem Ahmed for proofreading.

To anonymous peer reviewers.

Introduction

Evolutionary studies are no longer limited to biologists. Rather, as anticipated by E.O. Wilson and his vision of sociobiology in 1975, scholars in social sciences and especially in the arts and humanities have come more widely and systematically to embrace the synthesis of evolutionary biology and human culture. Thus, we can now write about the human mind and personality, morality, and the arts from a scientifically-infused perspective. In his book *Consilience*, Wilson notes how natural selection helps an organism survive and reproduce as necessary, nothing more. So the big question is not just how but why selection forces facilitated human beings to evolve minds, morality, culture, and the arts.

Why do we have culture when other intelligent species, without cultural achievement, have maximized their fitness to delay extinction and maintain survival? Our old mentality is nonverbal, dreamy, analogical and not coupled to analysis. Yet this infant thinking gets superseded by higher cognition and language. According to the laws of natural selection, we'd not have artistic or moral behaviors if they did not help us survive and reproduce.

Previously, studies in evolution and culture were separate and fragmented. In most of the generations preceding this writing, but for E.O. Wilson and his followers, the behaviorists, such as a Skinnerian, and their model of social construction prevailed. They claim, for instance, that the human mind is not adapted and instead the externality of culture makes one's mind. As David Buss tells us, whereas William James inaccurately emphasized and listed an incredible array of human instincts the behaviorists incorrectly reduced that list to one, our ability to learn, and so eliminated our human nature.

The seemingly disparate cultures of biology, psychology, social science, and the arts and humanities come together. We need to understand that the mind is adapted and that, like brain structures themselves, these disciplines work in association. So, too, human culture consists of intimately related parts invisibly working together, sometimes in unison, and at other times in conflict.

Questions about the distribution of wealth and resources or about solving the problems of poverty and education are not solely answered by banking or accounting procedures. Rather, higher order thinking that accommodates philosophical ideas would be responsible for addressing complex issues. Although we are born with an innate ability to learn, suggested by the ethologist Konrad Lorenz, and although we are acculturated into ideas as a species, 2 million years ago (mya) we were not, according to archaeologist Steven Mithen, so cognitively advanced. At the same time, some archaeologists, notably McBrearty and Brooks, find it implausible to accept any type of neural leap at

 | DOI 10.1163/9789004319486_002

about 50 thousand years ago (kya), since we were undoubtedly primed long before then for higher knowledge.

Were prehistoric hominins (ancestral, extinct, and modern human species), e.g., *Homo habilis*, simply apes with above average intelligence? What is intelligence? Wolfgang Köhler asked that question early in the twentieth century as he studied chimpanzees. Great apes are social, like us, but unlike apes we have foresight and planning concerning costs and benefits. Building from work by primatologists Jane Goodall and Köhler, one could define intelligence as non-routine responses. There is, says Goodall, modification of existing knowledge, multiple goals and even sub-goals. There are connections between knowledge and experience to solve a problem. More important than plans, there are moments of pause to deliberate among choices, a sign of reason. Intelligent behavior, and not necessarily instinct, is much more flexible and adaptive. An intelligent animal can refrain from acting on instinct and look for an alternative, which can be adaptive.

Liane Gabora and A. Russon ask what makes human intelligence. Aristotle says reason. Terrence Deacon says symbolism. Mithen says creativity. Others say theory of mind or guessing another's intentions. Recent work with great apes has blurred some human/nonhuman distinctions. Human beings evolved a detailed memory, attention, and ideas, which led in large populations to complex culture.

Natural Norms and Symbolism

Clearly there is something special about our species, but the special qualities of language and symbolic thought we possess were not bestowed upon us from a cloud-dwelling being. We evolved our higher order thinking skills in response to a number of natural elements and selection pressures. There were changing environments, the force of self-interest in larger group living, mentality in tandem with feelings and emotions, imitative behavior, the function of gesture and simple speech, cognitive mechanisms that enabled tool manufacture, and the development of social *mores*. Additionally, as biologist Richard Alexander suggests in *The Biology of Moral Systems* any *mores* are biological as much as cultural in that they have helped one group compete with another.

Within universals and mechanisms shared with other primates there are distinct cultural practices. Jane Goodall, in the conclusion to *The Chimpanzees of Gombe*, says that chimpanzees have symbolic thinking which probably arose with their need to understand how others and objects relate. While call signals, says Goodall, are emotional, intelligence is required to understand meaning, which most likely gave rise to symbolic thought. Higher cognitive abilities evolve out of complex social interactions, which require classification and definition, and fairly sophisticated tool use, which requires understanding of cause and effect. Goodall also reports that on three separate occasions young male chimpanzees were found practicing charging displays alone and in isolation, which indicates imaginary functions.

As organisms we evolved from biological survivors, to social cooperators, to symbolic communicators. Some organisms are short-lived and care less about survival and more about reproduction. Symbolism, which we have over other species, is in fact simply a sophisticated shorthand form of communication. We have facial expressions to convey universal emotions of happiness, sadness, surprise, anger, fear, guilt, embarrassment, shame, pride, compassion, and love. If emotions caused only irrational or harmful behaviors, natural selection would have eliminated them. There is, rather, a social, communicative aspect to emotions. We developed means of expressing emotions by manipulating materials around us, from cave art and ritual devices to, evident more simplistically today, digital emoticons.

The human mind, which deals with external phenomena internally represented, is supported by physiological brain processes, which are adaptations related to cognitive function and behavior. The evolution of the brain has much to do with our social intelligence, which in turn is linked to what we call the biology of morality as well as self-interested calculation. The long evolutionary environment of our early ancestors was social. It was in socializing, recognizing who is who and whether or not he is reliable, that broad personality traits developed. Behaviors related to social calculation and cooperation became values, practices, or beliefs and hence why I associate the rise of culture with flexible moral customs. Moral attitudes are variable precisely because of individual disagreement or cooperation with a group.

Being able to understand individuality and the specific strengths and weaknesses of others is related to adaptation. Openness, conscientiousness, extraversion, agreeableness, and neuroticism as well as their opposites reflect the social terrain in which we evolved with other people. In *Chimpanzee Politics* primatologist Frans de Waal says that chimpanzees have individualized faces and recognizably distinct traits such as walking, scratching, bodily movements, and sitting. Moreover, these apes are as socially responsive as human beings. The individual in relation to others in a group is fundamental to our primate heritage and accounts for our many social psychological mechanisms. These mechanisms can be, and often are, expressed as culture, e.g., sharing, singing, or playing.

In this book morality does not mean an inflexible code handed down from above but patterns of behavior that typically but not always integrate the individual into a community. There can be moral challenges to group status quo. It doesn't matter so much whether or not we approve or disapprove of some aspect of artistic culture. What really matters is that we react. Just as with evolution by natural selection, cultural evolution derives from individual difference, not just conformity. Artistic culture is graphic communication that holds values others can challenge. Stephen Davies sees our response to art culture as active and refers to art as a "social transaction" (52). I'd replace the word social with the word moral. I use the term *artistic culture* deliberately. Where the word art can be too narrow and the word culture can be too broad, artistic culture would include the values and beliefs assigned to the creation, acceptance, or rejection of art behavior.

Culture is a set of psychological adaptations for bonding, socializing, and cooperating and not simply instrumental. My point is that our approval or disapproval forms the basis of a moral judgment. In *The Artful Species*, philosopher Davies suggests that what is aesthetic and what is moral are distinguishable. I find that hard to grasp. There is little distance between viewer, creator, and object in what I say. In fact, I see immersion or conflict. Objects or behaviors are made, and they can be agreeable or not. If there is disinterest in art and aesthetics, then there is no material, practical, or ritual component to art, no function. If anything, we are completely interested in artistic culture.

Maybe Davies separates aesthetic and moral since he is light on continuities between apes and human beings. In her paper on primate facial expressions, Signe Preuschoft believes that any sorrowful display is evolutionarily bound up with caring. Dian Fossey witnessed a baby mountain gorilla actually cry after prolonged separation from its group. Faces and bodies engender emotions and mental representations. Approach or withdraw. Early versions of moral behavior, to be discussed in Chapter 2, evolve from mother-infant care and kin altruism. My point is that artistic culture reflects not only our inclination to cooperate with a group but more so to compete with the group or other groups. Artistic culture to uphold the status quo is either an offensive or retaliatory push against any who disagree.

Moral dimensions, moral scope, or moral making might even be better phrases since there is latitude in what we might term ethical (right/wrong) or moral (good/bad). Morality could focus on intentions or motives, outcomes or consequences, or social agreements. We can talk about morality philosophically and therefore value moral behavior. In nature, however, the value is not good or bad but based on survival and reproduction. Our human idea of value stems from the evolutionary process of enhancing fitness, principally on the individual and kin levels, but certainly on a group level. While fitness eventually deals with survival and reproduction, it includes other characteristics and abilities that contribute to survival and reproduction.

Many species engage in mutual benefit and reciprocal exchange. We see direct and indirect reciprocity quite often in primates. Most of the goods exchanged, however, consist of basic resources and mating privileges. We, on the other hand, have evolved more complex values associated with goods, so our exchanges are often multifaceted. In this way the evolution of our cultures has become moral since the evaluations of the goods and services can be related to beliefs. In some cases at a cost to us we will refuse an otherwise advantageous reciprocal exchange if it contradicts our personal or group values.

Elaborate artistic culture is like genetic offspring. Just as there are parental investments with children, it is so too with art and hence its moral components. Because parenting involves high cost we see why not everyone makes art since there is a loss of resources that could be self-allocated. So to make art the risk-to-benefit must be large. Art is genetically one's own and, done properly, it provides status longevity to the maker. The artist's investment in culture is moral since she knows the finished product could impinge on the values and beliefs of some others at a cost.

In the history of artistic culture we can see group norms of expression, ritual forms of expression, ideology, suppression of ideas, reformation, counter-reformation, and reformulations of all sorts. When I was drafting this book, we witnessed ancient art sculpture in Iraq being obliterated by the so-called fringe terrorist group ISIS. Culture and art behavior, in spite of many shared universals, is the privilege of an individual in a group. Artistic culture represents the exercise of the values and beliefs of a group and so shoulders moral weight.

While there are some moral universals, such as do no harm to kin, the larger social sense of right and wrong can vary from one cultural group to another, e.g. the use of force against foreigners. In other words, while there are social norms from one culture to another, these norms have been cultivated from more innate, natural tendencies regarding group loyalty and fairness. To emphasize, there are two layers. First, as in our primate cousins, there is the neurobiology of sympathy and emotions for kin. Second, there are the very large non-kin group level configurations of our species' moral emotions. In *Good Natured* Frans de Waal says that we are not born with "moral imperatives" (39) but do possess reasoning that helps us decide what the moral rules could be, though they are rarely fixed in time and place.

Artistic Culture and Moral Making

Was artistic culture more adaptive in the past than it is now? While material culture in our prehistory was mostly functional, such as body painting, tool use, and jewelry, it nonetheless carried symbolic and even moralistic messages. The body markings would distinguish one group from another. Historian and anthropologist Daniel Smail and Andrew Shryock say that body painting and tattoos are a form of social intelligence to affect another's behavior. That is moral. Forms of artistic culture, even now, can be termed moral since they can either uphold or challenge the status quo. Artistic culture is a psychological mechanism cued by the physical and social environments of sounds, shapes, and colors as well as interpersonal problems. Paleoanthropologist Robert Bednarik says that Paleolithic art enabled externalization of memory as *exograms* or symbolic devices for remembering. This culture is an output response to communicate emotions, ideas, status, individual and group identity, and specific problems or solutions.

Since these outputs encroach on others in some way, we can broadly term them as moral. Furthermore, borrowing from philosopher Denis Dutton, Anthony Lock, in this regard, has argued how abhorrent we find forgeries of symbols to which we have pledged allegiance. We have evolved a taboo response to important icons we find inauthentic. Forgery takes away from the identity of the creator and the forger, though we might actually admire the technical accomplishment of the forged product. Forgery, far beyond imitation, is the faking of someone else's character and life history. We value special creation, and our disapproval of integrity theft is a moral posture.

Oddly, though, completely overlooking theory of mind or the social brain hypothesis, Dutton insists that art has no moral or social import. Of course he'd say this with his insistence on art as sexual selection. Dutton sees art as a one-on-one transaction. This thinking ignores how some early forms of piercing, marking, and body painting were used to distinguish individuals in a group and between groups. Dutton almost seems to suggest art for art's sake with his emphasis, for the most part, on the Western tradition.

While I can grant the intellectual and emotional ingenuity of a maker to create spontaneously a special product, there is still what I call the peculiar morality of the artist. There is risk in the knowledge that implied viewers will variously interpret the created work and disagree in their approval/disapproval. As I argue in *Making Mind*, even art that disgusts us or reveals the evils of human nature is moral in how most people will react against representations of bad behavior. So Plato was right, in a way, to suggest that artworks as derivative imitations of Forms can mislead. Yet, we often learn what to do and how to act socially by witnessing examples of unsocial or immoral conduct.

Like our evolved taste for sugar and fat and the proliferation of fast foods, artistic culture is now rampant. We have an insatiable theory of mind, endless social communication, and continual questioning of norms. As community population increases, art cultures begin to differ dramatically, perhaps in response and as a solution to fragmentation. Artistic culture is an adaptation that *prevents* blinkered attention, avoidance of other ideas, or withdrawal from group activities. An adaptation can relate to reproduction through sexually mature mates who attract each other. Breaking narrow-mindedness, artistic culture is a reciprocal exchange which entails some costs and not pure altruism since culture can offend or even exclude people. Current examples of culture that have the power to influence other minds include food and fashion.

Artistic culture is predominantly an evolved psychological mechanism. The benefits to the creators of and participators in artistic culture include group identity, enhanced social status, mental play, signals of intelligence or desirability (sexual selection), or stimulants for cooperation. The neurobiological play of visual art could be a psychological byproduct of some other adaptive function. Costs include the time, energy, and resources to make, buy, or participate in art cultures. For early people, any risk in art making could have solved a problem of low status and competition.

Any form of artistic culture is not without detractors and does, therefore, have not only social but also moral implications. For instance, some artworks are politically controversial. Fashion models can be exploited by industries. Animal rights activists protest the use of leather and fur. Environmentalists and neo-socialists challenge any organizations, whether public, non-profit, or private that hoard and so deprive resources from others. From our prehistoric past, forms of artistic culture, too, were specifically designed to *cultivate* special trends, groups, and ideas, which would then exclude others. This activity is more than social and carries moral overtones.

Biologist Robert Trivers has demonstrated that group norms eventually extended outside the group. As a species, we typically avoid loneliness and gravitate to sociality. Our moral behaviors derive from this simple equation of connection, whether permanent or temporary, as do our culture and arts. Philosopher Patricia Churchland points out that like Confucius, Aristotle sees morality not as metaphysical but as practical. Our arts are physical manifestations of our practical need to consort with each other, especially when these arts are incarnations that challenge us to ask difficult questions about our humanity. Artistic culture espouses a natural morality in how it brings people together either in agreement or discord.

Social intelligence, which is strongly self-guided, exhibits itself as culture. Typically we are a caring and helping species, but we often merely suppress urges of anger and rage. Cultural glue and cultural division can be symbolically represented in art. We want to conform so as to fit in and survive. At times, however, any combination of emotions and higher intelligence can lead to the expression of something different or unusual. Any aspect of culture, such as visual art, is an adaptation by the intelligent, self-interested individual to operate successfully in terms of a group.

Tooby and DeVore describe an adaptation as a *compromise* between an individual organism and its situation, not only its habitat. They say that early human groups may not have had one set of behaviors but a range of complementary behavioral strategies. There is *no single process* of human evolution. A spectrum of possibilities includes increasing sophistication for sociality, communication, tool use, resource production and allocation, trade and exchange. We evolved culture to evolve us, and culture is a moral behavior.

Alexander in *The Biology of Moral Systems* and *How Did Humans Evolve?* would suggest that human moral issues are complex because of our tendencies both to cooperate and to compete within and between groups. We use our mental tools of consciousness, awareness, foresight, and planning, to compete. We are especially concerned with how we want to seem to appear to others. Although primatologist Dian Fossey shows us how the bond among mountain gorilla group members is so strong that they will fight off human predators to their own death, *we* would not necessarily call that behavior moral. Meantime, if a human parent were to break her bond with an infant by deliberately injuring it, we'd say that action is immoral. While I will use some form of the word *moral*, I am always using it in a natural sense. This book is not about great apes but about us. Yet we need to look at great apes to understand the origin of our own behaviors.

The focus of this book is ultimately on our species' expansion of *mores* in terms of individual self-interest against the group. We spend much of our time estimating the thoughts of others. Our brains did not simply enlarge because we began eating cooked meat and carbohydrates. Rather, as evolutionary psychologist Robin Dunbar might say, the brain became more flexibly modular and internally integrative because of the complex interaction of the individual in the group concerning motives, needs, and desires to obtain, butcher, and distribute meat. We spend a lot of time imagining what other people are imagining. With a bigger brain and more intelligence arose more social behaviors, culture, and art. What we call culture is a way for groups to cohere, to shrink out any who don't appear to belong or are not wanted. Even so, the origin of cultural norms can start with the idea or action of one innovative individual who is imitated.

Theory of mind is debatable in great apes (though see Miles) and is an often incorrect game of guessing another person's thoughts. But the fact that we engage in speculation reveals our ability not only to consider ideas on a higher level but also to attribute values and beliefs to others. With theory of mind, we know that others have motives, needs, and desires, and we assume a cause and effect relationship between intent and action. Infants only eleven months old can follow the gaze of others, and by four years children understand that others do not have to believe what they believe. Theory of mind is, then, central to cultural creations since it accounts for the making and viewing of, for instance, visual arts. We want, on some occasions, for others to know what we are thinking, and on other occasions we certainly want to know what another is thinking, so symbolic culture evidences a need. Frequently, though, we can only guess what someone else is thinking, which opens the way for calculating behavior.

In recent history, our higher order thinking was completely wrong, if one considers superstitions of the thirteenth century and witchcraft of the seventeenth century. In prehistory, Neanderthal brains were on average larger than ours, but yet they became extinct. In *Human Evolution*, Dunbar suggests that the visual area of the Neanderthal brain was large, due to the overcast environment in which they lived, but the important frontal brain for planning and large group socializing was small. Perhaps we killed the Neanderthals in taking over their territorial niche. Or maybe they died off since they lived in smaller groups and succumbed to diminished resources in an increasingly compromised climate. Maybe they became extinct because their brains were not as highly networked as ours.

On the one hand, we evolve culture, which consists of values, beliefs, and practices, and on the other hand culture has a direct impact on the particulate neural workings of our brains and so evolves us. What we call culture is essential to our survival. While there are certain foundations, in our symbolic language and group living, changes in culture reflect our brain's ability to adapt, its neuroplasticity, and the runaway effects any practice can have on behavior.

Human culture is not a free-floating enterprise but the product of social and sexual selection stemming from our prehistory. Culture has evolved through a course of variation and other dimensions of evolution. The behavioral areas of creativity and morality are related since they demonstrate the individual effort toward group integration. Part of the human evolutionary cycle of culture consists of values, beliefs, and practices as a cooperative effort, to say nothing of aggression, to satisfy individual needs. Not only are we biologically evolved creatures in brain and morphology, but our artistic and social behaviors, our mind and our manners, are also evolved in tandem to yield culture.

As Charles Darwin was perceptive enough to note, many animal species evince mental capacities that guide choices in terms of reproduction, what he called sexual selection. To an even greater degree seen nowhere else in the natural world, human beings exercise mental faculties, and have done so for millennia, that determine our physical, emotional, and psychological development as a species. Our sophisticated cognitive functions mark us as cultural creatures involved not only in sexual but also in social selection. In spite of our cross-cultural similarities which stem from our common evolutionary heritage, such as parental care, we nevertheless establish barriers to keep others away. For example, many cultures respect the notion of honor, but that code differs dramatically from one culture to another.

Emotions and Culture

What is culture? Is culture only a human production? Does what we call culture have evolutionary roots? What does culture have to do with moral behavior? Darwin saw other species engaging mental faculties in order to make reproductive choices. Human beings have evolved their brains in a complex web of external cause and internal need. We are highly effective cognitive mechanisms in terms of calculating another's behavior, predicting outcomes, and communicating with or deceiving others. Culture is not just a communicative signal but an integrative effect that carries emotional and moral overtones. Cultural manifestations in early humans, stone tools, operate with other forms of expression, like early speech. Culture is part of a group norm but stems from individual display.

Emotional and visceral responses work in conjunction with the beginnings of higher cortical activity and interaction. One can safely say that the older, limbic brain, which deals with response and life-sustaining mechanisms, is for self-interest whereas the prefrontal cortex is for, in most cases, consideration of others. The interplay and exchange, the cog of the neural firings in this network between these two brains, is perhaps what eventually yielded our symbolic culture. Processes of thought move in hierarchical ways, not linear. So an alert response is felt in the amygdala and then has to be processed upward associatively in other brain areas in order for it to be either ignored as inconsequential, quarantined for future consideration, or acted upon immediately, e.g., as fear.

Evolutionarily, emotions precede cognition. Emotions superordinate attention, motivation, and decisions. Gianluca Consoli says that our aesthetic emotions, which are shared, self-conscious, and creative, rose from individual and fixed primary emotions like fear, anger, disgust, sadness, and joy.

Although many species react emotionally without thinking, our private feelings and public emotions are intimately connected to our decision making. In *Descartes' Error*, neuroscientist Antonio Damasio says that of course emotions can interfere with reason, but most of our higher brain thinking is tinged with our limbic brain feelings anyway. Reactions are, in effect, added to themselves which create new cognitive emotions, asserts Damasio in *The Feeling of What Happens*. Our moral sense does not come from culture. It's more intimate than that. Moral sentiments are innate sensations and responses (via a so-called mammalian brain) mediated by our higher, evolved intelligence (the neocortex). There are multiple, interlocking mechanisms in the brain where feelings and thoughts can work in harmony or discord.

From an evolutionary perspective, according to primatologists Dorothy Cheney and Robert Seyfarth in *How Monkeys See the World*, cognition became an adaptive capacity for us to see relations between and among disparate bits of "information" (9). One might say that cognitive mechanisms evolve, not strictly behavior. Behavioral responses to a problem can vary, but there must be an underlying cognitive capacity to solve any problem. More crucially, our way of cognition changed dramatically about 1.8mya when we were able to imagine a symbolic self, that mental representation who can move and act in the past or future. The adaptive function of cognition in hominins is directly related to solving problems of what type of knowledge is needed to survive and how to behave among others. Advanced stone tools had already emerged by this time, which implies that our groups were interacting and sharing with each other. In fact, Sonia Harmand et al. demonstrate that diverse stone tool technology appears much earlier among hominins, at about 3.3mya.

Our bodies became small while our brains grew larger, suggesting less male-to-male physical competition and hence more pair bonding between sexes. Mental activity predominated over physical force. Charles Snowdon and Toni Ziegler say that the hormone prolactin, released in some monogamous female primates during nursing, has been related to oxytocin and implicated in fostering a pair bond. Both oxytocin and prolactin are pleasure and reward mechanisms for positive parental behavior.

The point is that fundamental to early cognitive and even morphological developments are types of moral components. And at the bottom of what we term moral are emotions or sentiments. Caring for and helping others, as well as deceiving others, would have been part of the fabric of an increasingly complex group structure. Moral behavior does not simply mean acting ethically; it implies, as well, breaking the norm. Likewise, an artistic creation or participation in culture is an expression of moral sentiments.

What do we mean by moral? We don't mean morality in its current incarnation related to set rules and codes. Rather, in its nascent form, what we now label as moral was a feeling, a sensation of whether or not a personal act or an action performed by someone else was harmful to another's survival or the well-being of a group. Primatologists like Dian Fossey, Frans de Waal, Biruté Galdikas, and Jane Goodall indicate that intelligent great apes, like us, behave with deliberation and calculation in groups. In his conclusion to *The Descent of Man* Darwin says any "moral being is one who is capable of reflecting on his past actions and their motives – of approving of some and disapproving of others" (680). This thinking, for us, often generates the creation of and participation in artistic culture.

Creative Powers of Mind

Beyond the obvious social implications of intersubjective reflection, any cognition tied to emotions can stimulate a symbolic action. We say that an aggressive chimpanzee *displays*, but it is unlikely that any display symbolically refers to an idea. Rather, being moral, in Darwin's terms, means literally demarking one's actions, and to do so one can elaborate an existing utile device or manufacture something purely for token display. For Darwin, intelligent and imaginary powers are necessary for any moral faculty. Unlike monkeys, chimpanzees can see themselves in a mirror and are therefore self-aware. Moreover, chimpanzees are capable of transferring and integrating information stimuli across brain modules, e.g., combining tactile and visual senses. This ability is an evolutionary adaptation since it provides the most information available about any object or event.

When an aggressive chimpanzee displays he signals. He does not refer to an abstraction. Our changing cultures imply that we don't simply signal but we use our intelligence in symbolic ways to refer to aspects of our sociality. It matters not whether those references are good or bad, right or wrong. Any creative connection to what is social we can term moral because it can impact how one acts or thinks about acting. Beauty and a sense of color are not limited to human beings. Apes are attracted to ripe fruit which has brilliant color. Male birds have evolved vivid colors and ornaments to attract females. We see among different species a preference for regularity, pattern, and symmetry. But the sense of what is beautiful in human beings differs because of the symbolic value it holds. We have moral symbolism.

In prehistory, there was no codified system of principles establishing good and bad or right and wrong. Certainly there were no ethics, but there were altruistic tendencies, initially between mother and offspring, later involving the father, and eventually spreading to others. Among mountain gorillas, Dian Fossey notes the extraordinary close emotional and physical bond between mother and infant in the months soon after birth. Biruté Galdikas tells us that an infant orangutan never leaves its mother during the first six months and is carried for up to four years. Early moral behavior would be the emotional response and expression of interacting with others, even if for one's personal gain as in the sustainability of genes.

While we are self-interested creatures, Frans de Waal has demonstrated that there is much empathy in our great ape cousins, a mental and emotional negotiation of what is pro-social and what is cunning. Any emotional activity would of course take place in the brain, and the more early hominins tempered their aggression or selfishness with consideration of others, the more often cooperative behavior cultivated. From there, cultural practices and beliefs spread, for culture consists of shared behavior and values. While any artist might create in solitude, an audience is always implied. The sharing and appreciation of something aesthetically special is social. And social behavior implies adhering to or questioning *mores*.

Cognition and emotions are the drivers of what we might call moral behavior. Much human conduct is cultural in that it cultivates symbolic thought and interpersonal interaction. Culture is an expression of emotions and cognition. Typically, most behaviors become accepted, copied, and passed along culturally. Jill Adams says the environment can alter molecules connected to an individual's genetic structure and so turn on/off or express genes affecting the body and behavior. Although he did not know gene theory, even Darwin intuited that the physical environment, e.g., light and temperature, had an impact on coloration and eyesight of the individual.

The human individual acts in a social context, and that behavior is a pivot of, to use social psychologist Jonathan Haidt's terminology, moral emotions. We could also use the term moral sensations, since there is a feeling of approval or not. We could also say social instinct. In *On the Origin of Species*, Chapter 7, Darwin terms as instinct when an organism is *impelled* like others of its species to act a certain way without experience, though some *judgment* is involved. Can we say, then, that there are cognitive emotions?

Yes, for these would be akin to our moral senses or those visceral responses that speed to and then ricochet off higher order thinking. An initial reactive moral sensation is reflexive, whereas conscience is reflective. The cortex is built on top of the mammalian brain. Antoine Bechara, Joseph LeDoux, and Antonio Damasio, to name only a few neuroscientists, have concluded that emotions play a key role in some decision making. Emotions are quick, unbidden motivators either to act or contemplate action. LeDoux asserts that brain connections from the emotional configuration are very strong. The mind evolved to find patterns and to make meaning, to accept but yet to overcome ambiguity. This process involves emotions and reason.

The Culturally Moral Mind

Culture, then, is the physical manifestation of our evolved mind in its attempt to shape that which is formless, to hammer meaning out of uncertainty, and to craft design from disorder. Thus we can see how emotion plays a part not only in artistic creation but also in aesthetic response. The creative powers of mind stem from interaction with a mixture of the physical environment and existing culture to release, essentially, the so-called selfish gene in its quest for transmission, longevity, and dominance. But here, dominance means control, and for the most part, mental control, which is not necessarily learned.

As psychologist Merlin Donald might say, artistic culture can control attention to network multiple minds. For human beings, as opposed to great apes, what we call artistic behavior is rehearsed and modeled, not routinely episodic. The brain has evolved to direct attention and to make judgments about information. In many ways the brain's ability for abstraction has evolved to make culture, i.e., to conjure symbols, manipulate representations, and evaluate actions and events. While a physical product appears, whether a cave painting or mythic hero in a story, artistic culture is a mental behavior.

Since we evolved in social groups, our minds evolved for survival and reproductive purposes to exploit others in complex ways that also evolved cooperative tendencies. If a behavior is social, involving actions and thoughts with others, it can be termed as moral behavior. Emotions are, Damasio would say in *Self Comes to Mind*, value-laden, from physical disgust to moral outrage. Clearly, then, cultural manifestations springing from evolved emotional responses tempered by and influencing cognition have a moral component.

Social behavior is broadly construed as relational group action. More specifically, the evolution of culture is a manifestation of sensations and emotions, the expression of moral feelings in social behavior. Everything comes down to the question of what one needs, wants, or desires. Sensations and sentiments like envy, jealousy, affection, and aggression all appear in our near cousins, great apes, and in our distant relations, Old World monkeys.

All forms of culture bear some mark of moral emotions, which we use to communicate directly or indirectly with each other. Each of us has a distinct temperament that will act and react among other individuals. We do not forage and survive alone. Thus, in the words of Frans de Waal, there is *chimpanzee politics* among *Homo sapiens* in how we act with self-interest or altruism, how any individual promotes himself or not. For wild chimpanzees, Goodall says that whatever seems to be done socially for the group, is no doubt driven by evolution and must be somehow advantageous to the individual.

While we can parse evolution and human culture into mind, biology and morality, cultural practices, and art and aesthetics, the defining thread that weaves them together is how our mental faculties evolved and adapted self-interest in relation to the group. Self-interest can become transformed into altruistic and creative, or sharing, behavior. The environment shapes only so much since there must be an innate capacity to shape. Eckart Voland and Karl Grammer say that our ability to detect beauty is innate. I would not disagree, but my emphasis is less on the detection and more on the response.

What does one do when feeling beauty? Beauty is not just physical but mental. We have pleasant attitudes and cooperative behaviors in response to what we find beautiful. Or not. Art and aesthetics are the results of genes reacting to stimuli from physical and social environments. Our moral sensations, in part both emotional and cognitive, are what propelled our advance over other species. We don't simply react instinctually, but instead we often think through our feelings and emotions. The results of thinking and the products of emotions create artistic culture.

CHAPTER 1

Prehistory and Mind

In *Self Comes to Mind* Antonio Damasio says the evolutionary advantage of consciousness is that it adapted human beings to have a theory of self, memory, reason, and of course language, all of which culminate in a flourishing and equally adaptive culture. The brain is a super organ that keeps the organism in homeostasis while monitoring the environment for dangers as well as for resource opportunities. Initially, homeostasis was for the body, but over time as our brains evolved in groups, social homeostasis became equally vital for survival. Old World monkeys, great apes, and human beings prefer to function intimately in groups as few as ten but no more than one hundred, and share extensive training of their young as well as adult social play. But the human brain, both a neural processor of connections and a symbol producer, is without question larger and more complex than any other. How and why did this occur? The brain is a survival mechanism, so the question remains as to how ours became reflective.

Interdisciplinary Nature of Evolutionary Studies

The volume by philosopher Gary Hatfield and archeologist Holly Pittman, *Evolution of Mind, Brain, and Culture*, offers perspectives on a number of interrelated hominin evolutionary subjects on the evolution and adaptation of the human mind.

While other species have debatable culture, like the potato washing of Japanese macaque monkeys, insect probes of chimpanzees, and singing of whales, human culture is obviously deeper and broader. This begs many questions as to origins. Since the mind is functional, what then are the evolved purposes of many cultural manifestations? There are cognitive capacities that arise from our inherited biology, and then there are those capacities that develop culturally. In his Introduction, Hatfield favors paleoanthropologist Richard Klein, providing an overview of parts of *The Human Career*. This is noteworthy since, later in the book, anthropologist April Nowell will favor the Sally McBrearty and Alison Brooks paper that questions the validity of Klein's so-called neural leap in *Homo sapiens* at 50kya.

Like Klein and others, the notion of branching derived from Darwin's single diagram in *Origin of Species* is a fundamental concept to human development,

 | DOI 10.1163/9789004319486_003

not the linear progression from crouching ape to upright modern humankind depicted by uninformed media. So we can see that there are similarities in human beings and other species going back a few hundred million years. Such related structures or homologies include, Hatfield notes, "photosensitive pigments in the eyes…frontal eye sockets…limb autonomy…grasping extremities… opposable thumb…" (7), to name a few. Homologies are similar in structure but not necessarily in function. The human brain shares structural similarities, e.g., the amygdala, and even basic functions, e.g., social communication, with other species, but on a superordinate level our brains function differently, e.g., theory of mind and mind sharing.

Certainly Hatfield and other authors are looking at a broad expanse of prehistory, so the individual is not really a factor. Thus, Hatfield can say with confidence that culture is less genetic and more environmental, that humankind creates its own environment and special niche. This thinking echoes the Lamarckian elements in later Darwin and recently the Darwinian model of cultural evolution. Lamarck emphasized the inheritance of acquired (not genetic) characteristics.

Perhaps the hallmark of being human is just the ability to evolve and develop through culture and not only through biology. There might be a trend that affects neural patterns, and that is biological. But the originating push to this trend might not have been biological but cognitive. There might have been selection for values and beliefs which would instance the dual evolution of genetic and cultural matter. Anthropologist Terrence Deacon, in *The Symbolic Species*, addresses co-evolution when he says that "behavioral adaptations tend to precede and condition the major biological changes evident in human evolution…" (345). Changes in behavior can be rapid and spread widely.

There are four types of social learning, according to Hatfield. First, in the words of psychologist Michael Tomasello, there is stimulus enhancement where one is cued to an outcome because of the environment. Second, there is emulation of another's behavior to achieve a similar outcome. Third, there can be exact copying of another's movement, which therefore involves motor imitation. Fourth, there can be an occasion where one intervenes actively to instruct another, meaning that intentionality is involved. Some nonhuman animals behave in ways to suggest social learning, and we see this especially in chimpanzees and vervet monkeys. However, stimulus response and imitation do not rule out any genetic bases to learn socially.

There is a marked difference when it comes to human beings and chimpanzees. Regarding "tool design," says Hatfield, chimpanzees are not capable of making or transmitting "improvements" (16). Whereas chimpanzees have a power grip, human beings have evolved a precision grip. Only human beings have evolved teaching and learning that expand from tools and hunting to the wider realm of culture in ever-extending groups. Groups then spin off satellites and variants of any preceding culture. Consider language. At one time there was a root language, but over time that has exploded into thousands, principally through group expansion and division.

While other species (chimpanzees and wolves) can form groups, they lack the complex, multi-layered cooperative, goal-centered abilities of human beings. Middle Pleistocene (780-127kya) Neanderthals and *Homo sapiens*, separately, engaged in highly coordinated game drives that presuppose some type of linguistic capability, which only followed from cooperative behaviors that had evolved much earlier, says Hatfield. Over the course of human brain evolution, the occipital lobe (visual processing) shrank in proportion to an increase in the parietal lobe (integration of senses). Over the past seven million years, according to Hatfield, chimpanzees have had more evolutionary change than human beings and so have probably lost more of whatever our shared common ancestor had. But unlike other species we have evolved language, as well as a rich symbolic internal and external life.

Nonetheless, Hatfield cites Seyfarth and Cheney who suggest that baboons are capable of "conceptual knowledge" stemming from social capacities which over time served as the basis for linguistic capabilities (24). In *Chimpanzee Politics* de Waal says that primates accumulate an enormous amount of information about conspecifics, particularly in attempting to anticipate the motives of others. So their intelligence is evolved from group living. Dian Fossey points out, for instance, that in terms of night nesting there is a degree of forethought since the mountain gorillas will choose places of best perspective to spot human poachers. In terms of human evolution, the team of Kaplan et al. claim high intelligence along with extended lifespan, childhood growth, grand-parenting and male provisioning co-evolved with the introduction of a nutritious diet.

Others in the debate of mind/brain evolution, Hatfield tells us, include Merlin Donald, who sees parallels among motor skills and representation. Donald says that mimesis is a means for one to abstract an object or event. Those like Tomasello lay importance on theory of mind. Philosopher Peter Carruthers sees creativity and the mental capacity to make inferences in terms of cooperation as a crucial adaptation. Mithen lays stress on Gardner's multiple *intelligences* which evolve into cognitive fluidity. These authors will be discussed.

Incidentally, Howard Gardner's intelligences, from the third edition of his book *Frames of Mind*, are Linguistic, Musical, Logical-Mathematical, Spatial, Bodily-Kinesthetic, and Personal. The last includes interpersonal, seeing another's intentions and motives, as well as intrapersonal, understanding one's own feelings and motives. Gardner has also discussed a Naturalist intelligence, regarding life forms, and an Existential intelligence, to pose ultimate questions.

While Gardner says that intelligences appear in everyone, some are certainly stronger than others. The point is that these intelligences helped us survive novel challenges, and so we did not have to be primed behaviorally to respond to unusual stimuli.

Other theories about human cognitive cores could include the essentially gene-based sociobiology of E.O. Wilson and the cultural evolution, akin to group selection, of social scientists Robert Boyd and Peter Richerson. There is also the evolutionary psychology of Leda Cosmides and John Tooby who look for explanations of human behavior not in genes or culture but in how brain modules adapted according to the environment of evolutionary adaptedness in the long Pleistocene, 2.5mya to 11kya. From a slightly different perspective is philosopher Kim Sterelny, who emphasizes human behavioral response to changing ecology. These authors, too, will be discussed.

Being Human and Language

Anthropologist Theodore Schurr asks, "When did we become human?" Researchers are beginning to conclude that traits like culture and tool use previously thought to be exclusively human are really "elaborations of similar features in other species…" (45). For instance, great apes have underlying capacities and abilities for language and symbolism, according to Schurr, and they are able to communicate vocally. Frans de Waal, Schurr notes, has drawn parallels between human language and the vocal and gestural manifestations of chimpanzees and bonobos. We see, too, not only complex social hierarchies in vervet monkeys, but vocalizations that identify rankings. Nevertheless, as Gary Marcus points out, language can be dissociated from cognition since it comes very early in life before mathematical reasoning.

The mirror neuron area is involved in language. Vocalization is rooted, still apparent, in gesture, and so narrative and even reading are akin to the doubling, as-if feedback we see in simulation theory and especially in mirror neurons. Paul Armstrong says language is a neurobiological social connector because it literally "reanimates" (158) us via visual and motor areas in the cortex. Language, in spite of what some would have us believe, is not necessarily the sine non qua of communication. A bodily expression or work of artistic culture can transmit more emotional information more quickly.

The language capacity probably came from a common ancestor to human beings and apes, but because the change in human beings came much later it has had a lasting genetic impact. Gabora and Russon, invoking the so-called Baldwin effect, say that the more often more people used language the more it was valued and consequently used. Those individuals whose genes precluded them from participating in this sequence were selected out. There is no sudden emergence of language nor a specialized gene or brain location just for human beings. Rather, a gradual confluence of adaptations regarding communication, tool use, mate selection, and sociality came into play.

Did the human brain simply evolve cognitively from sociality or for tool manufacture and use? We should not be surprised to see that our nearest cousins, chimpanzees, exhibit concerted, though primitive, tool use. Tool use is less likely with gorillas since they feed mostly on vegetation and are quite adept at manipulating leaves, stalks, and stems, says Fossey. Part of the puzzle with human evolution is that there is no linear descent, and even Darwin talks of branching and the imperfection of the fossil record. Schurr tells us there is a debatable hominin species, *Sahelanthropus tschadensis*, from 7mya, which shows evidence of bipedalism and so earlier than previously thought.

There is still much debate on the species that originates the *Homo* line. One might also refer to Antón et al. about the still open speciation questions regarding early *Homo*. Brian Villmoare et al. have also clouded early *Homo* origins by suggesting a predecessor from at least 2.8mya. Furthermore, recent finds by Lee Berger et al. discuss an undated but early *Homo naledi*, which Dirks et al. suggest might have deliberately buried its own dead. According to Schurr, *Australopithecus garhi* could be a key player since it might have connections to tool manufacture. New evidence suggests that *Homo habilis* and *Homo erectus* were quite distinct in spite of cranial similarity, but at any rate, many agree there was a decisive shift in *Homo ergaster* toward modern human form and behavior.

Simplifying, one could say that *Homo heidelbergensis* via *erectus* 600kya appears first in Africa and then spreads and gives rise to *Homo sapiens*. Most recent genetic evidence between Neanderthals and *Homo sapiens* (though both hominin) shows distinct variance at the species level. Neanderthals had very little breeding with *Homo sapiens* so no ancestral force on human development. Current technology suggests Neanderthal and *Homo sapiens* DNA sequence split 706kya to 500kya, long before the appearance of more modern humans about 200kya. Indeed, Aida Gómez-Robles et al., who analyzed dental morphology, suggest that neither *Homo heidelbergensis* nor *erectus* is the last common ancestor to Neanderthals and modern human beings. The divergence is older.

In the branching model, according to Schurr, features are "mixed and matched" (54), implying that forms and structures like bipedalism or brain size could have erupted and declined in any number of species. In other words, there might not be a way to find the so-called missing link. Without completely interpreting the statistic, Schurr says that the hominin population was significantly reduced after it split from the line that connected it to the chimpanzee. Is one inference that the trade-off or adaptive advantage was so great as to shrink the population? If so, what was the benefit gained? There might not be any way to pinpoint what the adaptive advantage was, but clearly the line from our early hominin ancestors in the australopithecines to later *Homo* species indicates a co-evolution of morphology and cognition.

Key genetic changes in the *Homo* lineage affected olfaction, bipedalism, diet, hearing, and vocalization. Schurr notes almost fifty genes implicated in olfaction have undergone change, and obviously we rely on the sense of smell very little. For bipedalism there were many physical changes from, to name only two, the pelvis and the lumbar. No matter when *Homo* began to be completely bipedal, the adaptations had been continual and significant. Dietary change, as per the savanna hypothesis, where fewer trees led to scavenging carcasses and to rudimentary hunting, led to more carnivory. Ultimately, in comparison with chimpanzees, human beings have a shorter large intestine and a longer small intestine since we more often digest meat.

Cause leads to effect and vice versa, so that increased carnivory could have led to stone manipulation which led to higher cognitive function. Other writers like Klein offer extensive detail about hunting, but Schurr's point seems to be that the dietary change of about 2.5mya might have led to more cooperation and coordinated efforts, not seen in chimpanzees, which could have prompted some form of communication. Of course it is at this time that early Stone Age tools begin to appear, presumably for butchering, though hunting is still unsophisticated. Could the quantity of tools have been produced by a species that did not communicate or cooperate with each other? Higher forms of collaboration would, in turn, increase brain activity.

Jane Goodall says that communication among chimpanzees is quite varied and includes many sounds, movements, touching, and facial expressions to indicate physiology, emotions, and rank. Any combination of communication signals is quite often used to affect another's behavior. There is a transfer of information via signaling and displays, much like a conversation, in order to predict another's behavior and so modify one's own. The key here is how these communication signals and behaviors are linked to social knowledge, i.e., the adherence to or deviation from norms. Although chimpanzees do not possess the physical characteristics to produce certain human vowel sounds, they can be taught sign language.

The FOXP2 gene (forkhead-box P2 transcription factor) is crucial to human speech production and comprehension and probably occurs in modern humans at approximately 200kya. But this estimate does not preclude other, older forms of gestural communication combined with single or paired utterances. Apparently, Schurr notes, this gene has a similar form, related to motor skills, in other vertebrates, but certainly one gene alone is not responsible for language. There might be some correlation between brain size and full language, but even small-brained species exhibit the ability to communicate with those they recognize, a social function. Not to push the point too far, but many species communicate vocally, and vervet monkeys utter certain grunts for specific social situations.

So there are genetic as well as cultural factors at play in early hominin information exchange and group correspondence. While many authors stress brain increase in cortical complexity, Schurr points out that in the human genome there are genes for microcephaly to restrict brain growth. There is a significant difference in white brain matter (not gray) between human beings and apes. The white matter, says Schurr, is evolutionarily significant regarding "connectional elaboration" (83). Schurr suggests that while some parts of the *Homo* brain increased or became more active, other parts lessened in importance, such as differences in masticatory genes between modern humans and apes. Lauren Gonzales et al. show how an Old World monkey endocast of 15mya reveals a modern cerebral complexity that suggests advanced organization separate and prior to encephalization.

Cheney and Seyfarth, in *How Monkeys See the World*, have examined the vocalizations of wild vervet monkeys. Vocalization for vervets is a form of thinking since it consists of a range of communication and information about predators. Vervets can respond to the alarm calls of other species, and while infant monkeys babble and are predisposed to call and signal they must learn particulars. While there is no hard evidence, most of this begs the question about whether or not monkeys can, like us, use signs to modify what another is thinking. Calls on one level are simply signals. Yet on some evolutionary level monkeys require representations in terms of social relationships.

Genes, the Human Brain, and Tools

Jody Hey employs genetics to investigate the origins of the modern human brain. On the one hand, human cognition and sociality differ in degree among people because of a genetic basis. On the other hand, says Hey, there are advanced capacities and sophisticated abilities that belong only to human beings. We share but then differentiate certain values and beliefs. The human brain, Hey notes, expresses up to seventy-five percent of every gene in some way, implying that human DNA has been adapted in terms of brain proteins. Since the human brain is a thinking mechanism, the emphasis here is on the cognitive aspects of culture, where culture is morally symbolic in its social capacity. Nonetheless, I am not discounting the enormous impact that first responders, emotions, play in our thinking and decision making.

Evidence, says Hey, suggests that human adaptations occurred between 6mya to 50kya, and then between 200kya to 50kya. However, Hey notes that sequencing human DNA is problematic since there is no indication of any gene's impact on "development," and even so the sequence concerning adaptation will probably not show "trait differences" before and after adaptation (98). Hey goes on to note that because ten percent of human "protein coding" genes experienced adaptation since the split from the human and chimpanzee ancestor, and since the human genome has approximately 25,000 protein coding genes, about 2,000 human genes have evolved over 6 million years. Moreover, says Hey, there are probably a number of advantageous modifications adapted for that will simply not be measured.

Take the FOXP2 gene which evolved gradually as an amino acid, implying that it had been present in some form for much of our prehistory, and then jumped to maturity around 200kya in the hominin branch. Compared with chimpanzees, according to Hey, this gene affects about one hundred additional genes.

William McGrew says that chimpanzees use tool sets where multiple tools are used in sequence to achieve one goal. There is also evidence that chimpanzees routinely use two or more tools simultaneously. This technological behavior indicates complex cognition. Concerning apes and monkeys, Seyfarth and Cheney discuss the primate mind before tools, language, or culture. While obvious, it is worth stating that the first occurrence of a trait or artifact is not necessarily the first instance. While basic human tools appear at about 2.5mya the mind and behavioral traits that produced those tools had been developing preadaptations. As Andrew Shryock, Thomas Trautmann, and Clive Gamble say, physical artifacts express cognition through objects. Similarly, in "The Primate Mind Before Tools," Seyfarth and Cheney try to look back to a hominin brain before cultural artifacts, when sociality and communication were emerging.

Both Seyfarth and Cheney work with baboons and other Old World monkeys who have many types of socially contextual vocalizations. The authors posit that there were important vocalizations before cultural language. Some of their research demonstrates, for instance, that baboons can form and understand intragroup storylines, which reveals cognition in tune with subtle and graded tones. Without language, animals are able to form perceptions of individuals from sounds and gestures, and Seyfarth and Cheney go on to suggest that the concept of personhood is therefore evolutionarily very old. If tool use is ultimately social, we see its roots in interpersonal group behavior.

De Waal, too, insists that chimpanzees have distinct personalities and can make individual discriminations, from intelligence to mood, crucial in terms of how any group is quite varied and competitive for status and attention. Seyfarth and Cheney suggest that this sense of personhood permits animals, particularly baboons, to posit theories about another's behavior. Baboons are able to *classify* each other through abstractions, as if each has an individual essence. While this mental activity is not symbolic culture per se, we can see here a form of abstract cognition which, in hominins, led to material advances.

Regarding tool use in chimpanzees, orangutans, and bearded capuchin monkeys, Kathelijne Koops makes the distinction between the opportunity and the necessity hypotheses. Along with co-authors Visalberghi and van Schaik, Koops claims that it is opportunity from the environment, seen in the abundance of ants, seeds, or nuts that fosters tool use. Koops makes clear, though, that there is a web of factors. The environment presents an opportunity, an individual will innovate, and then there can be social learning. Nevertheless, there must be, at base, a cognitive capacity to generate tool creation and use.

Prior to tool culture, natural selection worked to produce traits favoring sociality in advance of sophisticated behaviors. Seyfarth and Cheney indicate that baboon communication is related to rank, revealing unambiguous vocalizations in a hierarchy. Other vocalizations, Seyfarth and Cheney say, like predator warning calls, seem to be *evaluated* according to another's *identity* and motives, especially in terms of social relations. There's an association between linguistic utterance and cultural significance. Tool use is dependent on higher cognition and basic cooperation and intergroup communication.

Coordinating Cooperative Goals

As suggested by Wolfgang Köhler (and see Gómez), using an intentional gaze great apes know a human being can be a helping agent. Apes know that in a helping capacity the human agent can similarly perceive a problem and so share mental perceptions. On a fundamental level, moral sensations of approval/disapproval most likely arose between individuals, to judge and reward cooperators and to avoid and punish cheaters. Psychologist Felix Warneken focuses on the origins of cooperative behavior and finds that individual and collaborative helping are different forms of cooperation. Much depends on how well an individual can determine another's intentions, for without understanding intentions there is no personal motive or altruistic need to help. This type of mind reading became quite sophisticated.

Strictly speaking, cooperation involves agents who can respond to each other, who understand each other's goals, and who wish to share a common outcome, says Warneken. Infants as young as one year old are able to distinguish purpose from accident and can even estimate another's intention. In many controlled experiments, according to Warneken, infants spontaneously help adults though no reward is forthcoming. All similar experiments continue to reveal that there are innate helping and caring mechanisms that are cognitively primed and associated with other emotions and needs.

Furthermore, Warneken points out that research shows the willingness of young children to help achieve another's goal even when there is cost to themselves. Strategizing is a higher cognitive function. From an evolutionary perspective, similar experimental research with chimpanzees reveals that they are not completely self-interested and will act in altruistic ways when there is no clear benefit in sight. Here, too, we see cognitive mechanisms operating in association with moral senses. Perhaps this is not surprising given that the chimpanzees in question were raised by and often around human caretakers. Certainly it is worth noting that human beings and chimpanzees can interact on a basic intellectual level.

Warneken continues. Some wild-born chimpanzees living in a Ugandan sanctuary with very little human contact were tested. They too expressed helping behaviors even when no reward was expected, and in some cases they helped at increased cost to themselves. Moreover, an experiment involving non-kin chimpanzees shows that in the majority of instances a chimpanzee will engage in a novel action, e.g., unchaining a door so another can get food. While this might seem a strategic move, importantly, says Warneken, the chimpanzee who helped did not beg for food after the other received nourishment. Nonetheless, this mental strategy is most likely weighted toward future gain.

Warneken believes that altruism is not solely a human trait and that chimpanzees clearly exhibit "social-cognitive skills" in deciding whether or not to help (163). Collaborative activities visible in human infants are not just a human activity. While chimpanzees can act together, as in a hunt, it is unclear whether this activity is true collaboration for a shared goal or simply evidence of self-interest among others. Working with Michael Tomasello, Warneken questions the prevailing notion that altruism is a learned behavior since it is seen in infants and in chimpanzees long before the effects of socialization. So while cognitive functions come into play in a helping act, the originating force is certainly innate. Evidence of altruistic behavior early on and in chimpanzees indicates its evolved, biological foundation, later nurtured by social culture, says Warneken. Individual cognitive differences will of course cause different expressions of altruism.

I will deal with moral senses in Chapter 2. The point here is to demonstrate primate and human cognitive continuities, however tenuous, across the spectrum of language, tool use, and cooperation.

Mind Sharing and Language from Cooperation

Many of our cognitive functions have evolutionary roots, as evidenced in other primates. Likewise, our cultural manifestations are an advanced means of communicating our morality, the expression of good/bad, right/wrong, worth/worthless. Just as Old World monkeys and great apes can, to some minimal degree, share minds, we have evolved that functionality exponentially so that we can share value-laden symbolic creations across completely different groups. Mind sharing implies a single mind and infers the tendency, therefore, for mind singularity. One need not share his or her mind, and that too is a behavior that carries moral implications.

In addition to the benefits of predation protection and sharing, group behavior is not without costs, such as competition for resources and mates. But we see groups in all primates so benefits outweigh costs, and group living advanced cooperation. Robin Dunbar posits that as groups became larger and social networks became more complex, the brain evolved, and in our species language evolved to help us monitor what others said and did.

Merlin Donald, in *Origins of the Modern Mind* and elsewhere, offers along these lines mimesis theory, a cognitive adaptation from over 2mya. Originally related to tool manipulation this cognitive advance enabled what Donald calls mind sharing to form and eventually to flourish. Donald sees cognition as the main driver of human evolution. Mimesis enhances sociality and emotional expression. There is both group coherence and individual expression. Donald emphasizes that the evolution of human cognition would certainly not be limited to one cause. For the increasingly complex hominin brain there would be an array of associated adaptations stemming from the creation of self in relation to others in the group. The employment of skills, suggests Donald, would be tied to any number of complex events and factors connected to, for example, eating, mating, and community action.

Bonobos, chimpanzees, and gorillas do not have human cognitive symbolism or sophisticated group thinking, and these special qualities are connected to Donald's notion of a mimetic adaptation. Take, for instance, the sophisticated patterns on ostrich eggshells from at least 60kya, discussed by Jean-Pierre

Texier. By this time there is a sophisticated, intentional graphic communication system to render group culture and social norms. Imagine the incremental precedents that led up to this modern advance in messaging.

Hominins would not simply copy or produce and repeat a stimulus response. Instead, the mimetic act was a long process of learning and sharing abstract information. In terms of language, and contrary to any separate processing modules as per Jerry Fodor or Noam Chomsky, Donald says in his essay that neuroscience shows how during brain evolution any brain area might have been operating for various capacities so that hominin language turns out to be a "'kluge' cobbled together..." after other brain adaptations (176).

Chomsky is not necessarily evolutionary, as are Pinker and Deacon, but the notion of innate language ability is similar to how Konrad Lorenz posited learning as an adaptation of mind. Others who question Chomsky's approach, language as a mutation only in the human species, include Patricia Marks Greenfield and E. Sue Savage-Rumbaugh, whose research demonstrates a *Pan paniscus* (Kanzi the bonobo) who learned and invented a proto-grammar.

Much of the cognition of monkeys and apes evolved because of the complexities of sociability. Cheney and Seyfarth, in *How Monkeys See the World*, say that primates can apply social knowledge in different ways outside the social sphere. For example, chimpanzees mentally associate nests with specific individuals, and mountain gorillas can follow the trail of another who is not present, both of which illustrate mental denotation outside of domain specificity. Advanced mentality equates to theory of mind as well as attribution, hypothesis, cause and effect, deception, intelligence outside domain specificity, and an ability to extend intelligence. Since chimpanzees use tools they understand cause and effect. Chimpanzees can distinguish between animate and inanimate objects and among nonhuman primates are best at making and the most frequent users of implements. Tool use includes the manipulation of materials for threat or attack, hygiene, and procuring food, all of which require mental capacities that are both social and nonsocial, according to Cheney and Seyfarth.

Without a social skill like attribution or interpreting the meaning of someone's behavior, there would be no teaching, no information about danger, no deception, and no empathy, all of which chimpanzees have. Whereas monkeys are more interested in observing behavior, great apes are more interested in guessing how other minds make behavior.

Advanced language is a much later development, but Donald's point is that without language, and perhaps with only gesticulation and vocalization, there was significant communication. The ability to choose the best stones, the sharing of knapping and striking angles, and the sharpening of the blade edge were transmitted by a sophisticated mimesis for actor and receiver. Stone manufacturing abilities may turn out to be older than we realize, for cut marks in bones 3.4mya, around the time of *Australopithecus afarensis*, indicate stone edge use. Or stones might have been used to break open bones for marrow. Stone tools could come well before *Homo*, and among learning and sharing cultural skills, Donald's mimetic theory implies group cognition.

Cognitive scientist Peter Gärdenfors writes how cooperation plays a role in the evolution of language. Cooperation rather than gossip, sexual selection, or the cunning Machiavellian mind, was the factor driving any adaptation for language. For instance, indirect reciprocity (A helps B and then C helps A) implies communication, and for evolving hominins, the changing and challenging environment encouraged proto and then full language, says Gärdenfors. Of course cooperation can include deceiving others for self-gain. Cooperation is, therefore, closely related to the main point of this discussion concerning cognitive and cultural advances sparking from complex and contradictory moral emotions. Gärdenfors believes the duration of immaturity in both ape and human infants is crucial to the test and play functions of mind, cognitive developments related to mind reading and language.

Whereas other animals respond to drives (hunger) and instincts (storing food), human beings truly plan. This forward-looking capacity, according to Gärdenfors, is particularly linked to cooperative behaviors. Nonetheless, future strategizing can be selfish in terms of sexual politics. De Waal notes that each sex has different goals, which means different behaviors. In *Chimpanzee Politics* he says that where the female chimpanzee is typically caring and offspring-devoted, the male is calculating and status-centered. Some of this behavior is certainly driven by genes, but much of it is culturally learned. Cognitive adaptations related to long-term planning include *intersubjectivity*, a term preferred by some over theory of mind since the former denotes shared thinking where the latter only implies mental guesswork. For the most part, then, much of our cognitive work is technically solipsistic, but not to say completely self-directed.

We cannot only guess another's intention but we can also share intentions, and this mental cooperation, from a very early age, is what constitutes human goal-making. Gärdenfors says much cooperative behavior goes back to *Homo habilis*, who with its small body relied on joint effort in scavenging, which also enhanced one's reputation in the group. With group efforts hominins moved from mere signaling to symbolism, which is internal and representational, so that language evolved to enhance cooperation. Increasing mental representation and meta-representation eventually leads to the fluidity of brain modules, as posited by Steven Mithen.

Cognitive Fluidity and a Shifting Timeline

In his essay, "Cathedral Model," Mithen revisits his theories of cognitive fluidity from his book *The Prehistory of the Mind.* Drawing from Gardner's intelligences and Cosmides and Tooby's modules, Mithen postulates that separate brain functions began to communicate with each other. For instance, sociality, hunting, and tool making eventually began to shift from domain specific rigidity, as in hunt equals food, to flexibility, as in hunt can equal bone which can translate to tool. Better yet, hunt can equal bone which can translate to personal adornment. While Mithen convincingly limits any cognitive fluidity in *Homo ergaster* and nearly negates any in Neanderthals, he labels McBrearty and Brooks' paper as "controversial." Mithen sees no cognitive fluidity among Neanderthal intelligences, so that decorative beads cannot flow from social or technical intelligence, spears cannot flow from natural or technical intelligence, and art cannot flow from social or natural intelligence.

Mithen sees full consciousness and cognitive fluidity as a very late post-Neanderthal development. One can see how Mithen would debate McBrearty and Brooks who, like April Nowell, marshal quite a bit of physical evidence to claim that Mithen and others (e.g., Klein) are wrong in suggesting a late, sudden shift in hominin brain organization. McBrearty and Brooks propose that some modern social and cognitive behaviors go back as far as 300kya and include Neanderthals. More importantly for this discussion, Constantine Sedikides et al. have posited that the symbolic self, a crucial turning point in hominin cognition, dates back 1.7mya. Terrence Deacon, in "Aesthetic Faculty," has gone as far as saying that a nascent symbolic proficiency appeared as early as 2.4mya.

In his "Cathedral" essay Mithen slightly revises his chronology, looking at cognitive fluidity appearing around 70kya since that is a firm date for the second out-of-Africa migration, the first at about 1.7mya, and for sustained and concrete symbolic behavior.

Mithen boils down evolution of mind to three theories. First, and quite convincing, there is the social brain hypothesis of Dunbar, where gossip plays a key role in the development of language and where theory of mind and social interactions foster complex relationships. As Dunbar himself says, because the human brain is so large it cannot simply have evolved, as it did in other animals, to survey and process information about the environment. In other words, says Dunbar in "Social Brain," information is not just stored but *manipu lated* since one knows she knows.

These ideas are crucial to Mithen since they elaborate his social intelligence which is important for both domain specificity and, later, cognitive fluidity.

The social brain hypothesis does not account for technical improvisations, like tool use, or explain why at approximately 70kya there is symbolic culture, even though modern brain size appears at 250kya. Second, there are Thomas Wynn and Frederick Coolidge, who claim that modern humans, not Neanderthals, have greater short term memory. This idea also supports Mithen and cultural developments in the Middle Stone Age with cognitive flexibility. Third, Geoffrey Miller emphasizes sexual selection via Darwin as the driving force behind mind and creativity but without integrating solid reference to the archaeological record and with no chronology.

For Mithen, continuing with his "Cathedral" essay, the FOXP2 gene mutation affects other genes implicit in language in anatomically modern humans (AMHs) bridging language, cognition, and behavior. Mithen does not believe there could have been a proto-language (words without grammar). He prefers a more holistic approach with the leveling of a series of *fixed phrases* related to singing since tonality would add subtlety and differentiation. Cheney and Seyfarth, in *How Monkeys See the World*, for example, devote considerable effort delineating the subtle differences among warning calls and social grunts of vervet monkeys. Early speech acts carried communicative or social import and were therefore foundational vehicles of moral emotions. Importantly, for early human beings any vocalizations would be representational and symbolically referent. We see, then, the building blocks of our artistic culture in non-human primates.

Mithen perceives any artifactual culture as potentially responsible for instigating cognitive fluidity and certainly for promoting and extending mind sharing. Artifacts are an extension of advances in ideation that had already taken place in the interaction between the limbic brain and the ever-growing neocortex, especially in a social context.

April Nowell examines the cognitive and behavioral evidence from the Middle and Early Upper Paleolithic eras to place advanced thinking well before that posited by Mithen. For Nowell, cognitive means "phonemic, syntactical, and symbol based language…" as well as the understanding of oneself, theory of mind, memory, and forethought (235). Combinations of these elements appear in early hominins based on factors such as, but not limited to, tool manipulation and the first exodus out of Africa. Clearly these are adaptive advantages, and Nowell, drawing heavily from McBrearty and Brooks, focuses on Neanderthals and AMHs. She defines and contextualizes what constitutes modern behaviors: "abstract thinking…planning depth…behavioral, economic, and technical innovations…symbolic behavior" (236).

Depending on whom one reads, the cognitive timeline shifts, and much depends on definitions of accepted expressions like thinking, understanding, planning, and self. See, for example, Matthias Meyer who questions modern human and Neanderthal origins as originating from *Homo heidelbergensis*. He establishers a date of archaic human origins in an earlier split from the shared *Homo neanderthalensis* branch, perhaps as much as 750kya. That predates conventional wisdom about our split from *Homo heidelbergensis*. At any rate, nothing is linear. There is an ebb and flow of adaptations, traits, and behaviors from various species that rise and fall and at times come together.

Ecology, Technology, Social Organization, and Symbolic Behavior

If genes flow two ways, why did Neanderthals not benefit from AMHs? Without directly saying so, Nowell seems to suggest, but does not assert, that the cognitive differences between Neanderthals and AMHs is exaggerated. The differences might have been more cultural with obviously fatal consequences for the Neanderthals. To build a case regarding the similarity between Neanderthals and AMHs, Nowell uses a template from McBrearty and Brooks that covers ecology, technology, social organization, and symbolic behavior.

This outline is worth reviewing since it demonstrates how a species closely related to but not exactly *Homo sapiens* expressed similar thinking and behavior. This contiguity is understandable but worth examining because the further step considers the continuities between *Homo sapiens* and other primate species. Yet it is important to stress in line with Darwin that every species is unique, with its own morphology and behavior, developed over a long period of time.

Ecological evidence, says Nowell, suggests both AMHs and Neanderthals were involved in big game hunting. Both also cooked and ate small game as well as a variety of other products from mollusks and turtles to olives and pistachios. However, not all Neanderthals ate coastal yields, an important food source for development and survival, and AMHs overall ate a much wider variety of small animals whereas Neanderthals were geographically limited. Even if we agree that Neanderthals are a separate species, there is no reason to doubt, as Nowell suggests, that they were cognitively advanced.

Concerning technology, Howiesons Poort, S.A. has the first truly Middle Stone Age artifacts, says Nowell, but recent analysis questions whether these people were AMHs. Further complicating the timeline, Nowell says other areas of Africa north and south of the Congo reveal "elements of modern behavior" though these AMHs retain "archaic" features (244). Here then is a cognitive rather than a morphological advance. So it seems, according to Nowell, there are "pulses of modern behavior" that rise and then cease and then rise again.

Through McBrearty and Brooks, Nowell seems to want to blur the line between what we call a Neanderthal and an AMH, a view that is pretty much accepted here. Any manifestation of modern behavior implies even earlier advances in mentality.

For instance, Nowell says that Middle Paleolithic and Levant sites used by both Neanderthals and AMHs are difficult "to distinguish." Furthermore, any aim to classify tool industries between these groups, and the Middle from the Upper Paleolithic, is "speculative" (244). Steven Kuhn, Stiner, Reese, and Gülec argue that it is not accurate to say earliest Upper Paleolithic artifacts came only from Africa or central Europe. They appear in the Levant and were part of shared communication.

In Europe, too, there is evidence of Neanderthals involved in *transitional* industries. Thus there is debate about Châtelperronian and Aurignacian interaction, the latter ultimately intruding into Western Europe and usually ranked with those AMHs who later populated Europe. Still open is whether or not the Châtelperronian people moved by Neanderthals into the Upper Paleolithic, implying higher cognition in the Neanderthals, or were heavily influenced by AMHs.

A transition is seen, too, says Nowell, in southern Italy where Neanderthal artifacts seem to blend into the Upper Paleolithic products of the Aurignacian culture. Even at a German site there is evidence of Neanderthals making pitch adhesive 80kya under temperature controlled conditions, which also implies high cognitive function and communication. Keep in mind that sophisticated behavior implies a longer history of accumulating interactive mental functions. In terms of evidence for economic and social organization, Nowell tells us that Neanderthals occupied, left, and then moved back to certain sites, one in Germany dating to 120kya. Home site behavior is modern and controversial. According to Klein, any home base thesis would put many human behaviors, e.g. food sharing and division of labor, at 2.5mya.

Davorka Radovčić et al. examined eight talons at a Neanderthal site in Croatia and conclude there was symbolic behavior in this jewelry dated to at least 130kya. The authors are convinced these artifacts were neither made by nor copied from modern human beings. Concerning symbolic behavior there is evidence of burial rituals, including beads and skull marking/de-fleshing, in Neanderthals and AMHs from South Africa, Ethiopia, and Croatia as far back as 150kya. Beads and pendants bear drill markings; and it is possible shells were previously collected and shared. Ochre colorants for purposes like tanning hides and medicinal use clearly predate any revolutionary neural change of approximately 50kya. Paleoanthropologist Robert Bednarik says bodily adornment, or self-awareness shared among hominids (extinct and modern great apes) and hominins, probably dates back at least 2mya, so there is no punctuated, cognitive revolution overnight at 50kya.

There is evidence that ochre was shaped into crayon-like drawing pencils at least 60kya, as well as evidence of pigment manufacture and color enhancement in and among Neanderthals. In Africa at Twin Rivers, Nowell says pigments and colors can be traced back as far as 260kya, which predates AMHS. Sophisticated use of materials points to a physical manifestation of higher order thinking that had been developing incrementally. In addition to extensive use of red ochre, black manganese was also used.

As is true in much human behavior, the adaptive advantage is, broadly, social. There is gain in status, group identification, or mate selection. Nowell points out ochre compounds and forms of personal adornment have been found in Blombos cave, S.A., dating from between 100-70kya, and beads in Israel have been dated at 135kya. Use of red ochre dates to at least 300kya in India and the Czech Republic where it is not naturally occurring, implying that it was transported there. Evidence of personal adornment among early Neanderthals in Europe predates AMHS, but use and practice varied geographically. In a Darwinian sense the physical artifacts reflect the gradual increase in brain size and mental capability.

The upshot of Nowell's work is that Mithen's assertion of Neanderthal cognitive constraint, i.e. domain specificity without cognitive fluidity of later people, is off the mark. Nowell's argument is in light of all the factual evidence now assembled. What seems like a cultural explosion at 50kya might be the result of a series of contractions from the eruption of Mt. Toba, at 71kya, and glaciation for up to ten thousand years, forcing a sudden drop in cultural production and advance, to be picked up later. Movement of early AMH populations meant distinguishing themselves in some cultural way.

So while there is a cognitive process involved in symbolic marking, technically there might not have been a neural change in this development.

The Question of Micro/Macro Brain Evolution

Although most of our evolution seems driven by cognitive causes and effects regarding, for example, sociality and sexual selection, even Darwin would point to how the external world impacts adaptation. The scientific team of Peter Richerson and Robert Boyd, who cover culture/gene co-evolution in *Not By Genes Alone*, write about other non-genetic subjects in their essay, discussed here, "Rethinking Paleoanthropology." They ask why, considering its importance in so many aspects of human life, the brain did not evolve earlier?

The short answer has something to do with two types of evolutionary forces: micro evolution (small and fast) and macro evolution (large and slow). The changes within mammal brains, say Richerson and Boyd, have been occurring for 65 million years while undergoing variation due to climate changes. A secondary question according to these authors concerns the types of "processes" affecting evolutionary change and on which particular "time scales" (266). In other words, Richerson and Boyd suggest that not all emphasis should be on internal, micro, slow evolution.

There might be other reasons and causes to explain some rather sudden developments. Critical to this discussion, there is the question of *when* selection seems to favor a development, such as a larger brain. In the hominin line, not factoring in migrations and climate changes, the increase in complexity of group size plays a part. On scale, all vertebrate brains are similar, and nature would prefer a smaller brain since it grows faster and takes much less energy to maintain. Many mammals have experienced recent brain growth, so there is probably some "common external environmental change…" catalyzing this increase (267).

Of course social learning and genetics explain brain growth, and studies demonstrate, say Richerson and Boyd, that there is a balance between "innovation and brain size" (272). Fossil evidence suggests that hominins were rarer in the Pleistocene than other species, perhaps as few as several thousand in Africa and one hundred thousand in South Asia. In Swartkrans, S.A. at 1.8mya, these authors note, there are only three hominin fossils compared with thirty-nine australopithecines, thirty-three monkeys, ten large cats, thirteen hyenas, and fifteen canidae. How, then, did hominins survive and *Homo sapiens* flourish? The answer is that there was a co-evolution of both brain and culture impacted by external macro elements.

In their essay, Richerson and Boyd correlate brain/culture development with climate changes. During the age of mammals, from 65mya onward, there have been a series of different climates, with the trend increasingly cooler and drier, so that by about 3mya there are ice ages. As already noted, stone tools appear at 2.5mya with brain encephalization shortly thereafter. Then from 1mya, these authors tell us, there were one hundred one-thousand-year long "glacial cycles" related somewhat to the earth's axis wobble but more so to the "ocean-atmospheric-ice sheet" system (275). These developments were in turn connected to the range of the sea floor via continental shift and splurge of mountains, all of which affected wind and ocean currents, and ultimately affected biotic life.

Richerson and Boyd say there was a sudden shift 14,650ya from cold to interglacial warmth, and data seem to reveal that while glacial periods have flexible weather interglacial periods do not. They are clear that these extreme environmental variations correlate to their theory of brain development. Contrary to Klein and Mithen, they take an externalist and not an internalist approach and so, akin to McBrearty and Brooks, they challenge prevailing views about human evolution by imagining modern cognition and behavior well before the usual 50kya turning point. This means that Richerson and Boyd see the Upper Paleolithic developments in technology as cultural and not biological, not genetic.

Neanderthals were adapted to the cold, but AMHs were not, so the climate changes might have driven the cultural changes for AMHs whose bodies were more adaptable than Neanderthals. A glaring example of what Richerson and Boyd suggest is how in the Holocene, with more warmth conducive to plant growth, we see agriculture in many diverse populations. Pollen in earth cores and fossils of plant-eating beetles reveal a changing and challenging environment to the then (not now/Holocene) hunters and gatherers. These climate changes undoubtedly "imposed novel constraints…" (286) concerning how people adapted.

Richerson and Boyd go on to say that climate shifts also included drier, longer summers when there were more grazing animals where there might now be tundra. In the ice age glaciation there were fewer trees and more low level plants, generally. Some adaptive changes would include the ability to handle various hazards and to sustain any cultural innovations under climate pressures. If the population decreases, there is a statistical likelihood that those people best at making tools and other innovations will be lost. See, say Richerson and Boyd, how the Tasmanian toolkit was greatly diminished when the sea separated these people from Australia and the population dwindled.

Richerson and Boyd conclude, then, that toolkit sophistication might have as much, or more, to do with climate and demographics than cognitive advances. Susan Antón et al. in a similar vein suggest that early *Homo*, of which there are many but especially *Homo erectus*, evolved in "habitat unpredictability" where the savanna hypothesis has been over emphasized to the exclusion of more diverse vegetation and wetter areas.

Human Behavioral Ecology and Multiple Adaptations

Building from this recent work by Richerson and Boyd, Kim Sterelny examines, in a related manner, Human Behavioral Ecology (HBE). Sterelny emphasizes "the adaptive flexibility of human behavior…" so that "our learning and decision mechanisms equip us to respond successfully to a broad range of environments" (303). An example would be the adoption of agriculture over foraging. Sterelny notes how evolutionary psychologists are not fond of HBE because there is the question of whether or not "current behavior is adaptive." Evolutionary psychologists are more to cognitive adaptation in the prehistoric past. What then of the timing of birthing decisions in comparison to food resources, Sterelny asks? In other words, there is a connection between fitness and a complex organism's capacity to act adaptively, to take advantage of environments and circumstances.

But HBE is not without criticism. Sterelny himself says, for instance, in some cases there is the implication, especially for human evolution, that unless we change in accordance with a changing environment we suffer. This attitude overlooks how hominins have had the ability to shape their own environments in spite of external changes forced upon them. Sterelny's point is that over many thousands of years, *Homo sapiens* have clearly *optimized* resources and circumstances across environments. There were no limits on hominin cognition, resulting in multiple cultural creations in response to varying environments. This might help explain why Neanderthals and their hominin brain became extinct.

Likewise, Peter Carruthers sees "multiple adaptations" as responsible for a cumulative culture. He lists eight factors or "systems" working independently, though he is not really concerned with any chronology of cognitive adaptations. Drawing from the module model of evolutionary psychologists, Carruthers shows that there were a number of different adaptations and not just one or two faculties. This is an interesting approach that moves away from anything linear and even branching, suggesting instead a constellation of moving objects and forces. Carruthers is more in line with Tomasello's ratchet effect of learning and innovation, concerning the expansive growth of culture, but he sees these developments arising from many adaptations. The eight adaptive systems are as follows and worth noting.

1. Imitation, shared intentionality, and mind reading. Carruthers cites McBrearty and Brooks in terms of a longer and more gradual cultural advance emanating from Africa where there might have been, at times, cultural stasis because of low population or localized usage. Consider, for example, the so-called population bottleneck of 70kya consisting of about only ten thousand AMHs. Compared with chimpanzees, human infants share intentionality, which establishes an important "developmental learning sequence…" leading to imitative behaviors, skill acquisition, and theory of mind (330).

2. Language. Carruthers says language is less cultural (imitative) and more of an adaptation. While word learning is imitative, context driven, and involves understanding another's intention, syntax cannot be explained so easily and is therefore an adaptive mechanism.
3. Creativity. Culture would never truly advance without the ability to tackle and solve the many problems, social and technical, hominins experienced. Here, the evolved mechanism concerns "a disposition to activate and rehearse combinations of action schemata creatively..." (335). Nevertheless, even so-called technical problems have social implications, such as how a new tool or method can enhance one's status or help the group.
4. Fine motor coordination. In spite of an elephant's ability to retrieve one peanut with its trunk or a hummingbird's ability to balance while suspended in mid-air (his examples), only human beings have a wide "*range* of precise physical skills..." (336). Citing Merlin Donald, Carruthers says that only human beings "practice and rehearse" so as to make "skill-acquisition...an intrinsic goal" (337). Moreover, many people enjoy watching others exhibit and excel at skills, whether an athletic competition or an art exhibit. Social participatory experience bears, as I am using the term here, moral components of approval or disapproval. While one can appreciate the skills of boxers, one might not approve of the practice of two people deliberately punching each other to inflict injury.
5. Physics. More than perceiving cause and effect, grasped by other animals, we truly *understand* the powers of cause and effect, which gave only us the ability to manufacture and then improve upon stone tools. There is evidence of a human understanding of forces, like gravity, at as early as three months of age, says Carruthers. True enough, but Schopenhauer, in volume one of *The World as Will and Representation* (book one, Section 6), notes how an elephant *understands* that some bridge structures will not bear its weight. In his experiments in the early twentieth century, Wolfgang Köhler demonstrated that chimpanzees have mental representations. In one experiment the box used to stand on to obtain food was out of sight yet an object the chimpanzee had earlier encountered.
6. Inference to the best explanation. Even before our ability to create and prior to our real understanding of physics, through learning, imitation, and cultural transmission we became expert trackers and hunters (339). But here, too, we are not the only species who are good hunters. Perhaps what Carruthers means to say is that we evolved reason whereas other species still rely on instinct.
7. Cooperation and norms. These are probably very old adaptations in terms of hunting and scavenging. Every human culture has basic laws, expectations,

and permissible (or not) behaviors. A sense of fairness is universal, as is moral aggression, punishment, and guilt. De Waal is keen to note in *Chimpanzee Politics* that male chimpanzees seem to regulate their aggressive tendencies. Control is necessary since coalition formation and status-maneuvering within the group are more mental than physical. Dian Fossey on a number of occasions points out the high degree of restraint employed by powerful male silverback mountain gorillas. For our purposes, this adaptation of norms is an integrative adaptive system, and intimately related to the next one, that pulls in cognitive functions, moral emotions, self-interest, and group identity.

8. Further adaptation to culture. Individual learning is difficult and there is a myth of the lone genius. Rather, conformity is tied into the previous adaptive systems and accounts for individual acculturation. Conformity can help a group cohere and establish an identity while increasing variation from another group. Carruthers suggests there is an adaptive mechanism that *computes* behavioral differences and then decides whether or not to act similarly. This is an adaptation about copying whatever works well or is patently successful. This mechanism, too, is crucial in understanding the larger picture of moral sensations, emotions, self-interest, and group culture. Consider the artist who resists some acculturation since she has a different vision of the world. Some cultural conduct is about not conforming.

There are both cognitive (computational) and moral (emotional) responses for behavior. Of course there can be maladaptations. For example, our psychology evolved in an environment that prized status for reproductive success. Now, we perform and invest in activities for status that harm us, like overworking ourselves and conspicuous consumption.

Beyond Memes to Social Agreements

Quite pertinent, archaeologist Philip Chase claims that human culture is more than memes (social ideas) and simple transmission. Rather, culture stems from *concurrence* or an "ability and willingness to create and to abide by agreements…" in a degree of social life not seen in other species (349). In considering agreements, we come back not only to speech and the origin of language but to moral considerations.

Chase seems to lean toward social constructivism with his insistence that cultural variation is independent of genes and is more dependent on "social learning." Although he positively cites social construction, his point nonetheless seems to be that the social transmission model is "incomplete," placing culture, instead, "in the mind" and not in behavior (351). In this picture, we can see that both cognition and moral emotions are responsible for the rise of human culture.

According to Chase, a "state of mind" and "intentions" are expressed and shared, and that is what instigates behavior (352). This is when and where shared emotions and theory of mind factor into behavior. In this way there is an "intention pool" that is shared, altered, and even improved via *concurrence* or an agreement to share. However, this agreement does not strictly belong to those who share intentions. A mental settlement is on another social plane and cannot be completely changed but only *negotiated* into a new agreement by the two people, which can occur on a group level. Therefore, from mental activity arise norms of behavior and shared values and beliefs.

We can see, says Chase, how societies are "invariably organized by means of cultural concurrences..." (362). Biased toward helping and caring, people more often than not agree to agree. While human beings experience animal-like drives (hunger), unlike animals a cultural concurrence would consist of a sophisticated and goal-oriented game drive. Furthermore, cultural concurrences like values and beliefs can originate motives. Chase admits, though, that it is difficult to find evidence for the origins of concurrences, such as captive chimpanzees sharing human signs with other chimpanzees.

Some species seem ignorant, says Chase, that any other might have mentality. But work by Cheney and Seyfarth, worth noting again, indicates that in predator alarming and social grunting there are elements of shared mental representation. There seems to be evidence, for human beings, that covenants were an important factor up to 127kya, at which time brain endocasts (impression of cranial cavity) suggest a vocal tract. Brain size might have more to do with reorganization of structure than with a higher functionality like language, since in living human beings, brain size can vary greatly yet not affect language capability.

There is evidence, according to Chase, that brain reorganization started in australopithecines. Then there was an enlarged Broca area (linked to speech) in *Homo habilis*. Subsequently, brain asymmetry first appears in *Homo ergaster*. The lowering of the larynx is controversial, but speech adaptation would have come well before active language. Chase goes on to cite a number of examples of early artifactual, symbolic representation over various sites from Europe, the Mediterranean, and Africa, with a questionably human-like shaped quartzite from perhaps 500kya.

The evolution of mind, brain, and culture was not simply genetic but occurred according to response variations in any number of ecological and social situations. Since these adaptations were selected for and provided survival and reproductive benefit, they became genetic. I'd venture to say that artistic culture itself is a social agreement or a challenge to any perceived agreement.

From a Cognitive Animal to a Thinking Species

Michael Tomasello tackles cognitive evolution in *A Natural History of Human Thinking*. By his own admission, Tomasello calls *Human Thinking* a sequel to his *The Cultural Origins of Human Cognition*, though emphasis here will be on the later book which charts the evolution of primate cognition and human thinking in terms of cooperation.

A genealogy is important to this discussion since it highlights my claim that cultural creations are in fact elements of the individual assertion in spite of cooperative behavior. Because of our cognitive emotions we can like/dislike, approach/retreat. Physical artifacts and visual art serve as a means of this cooperative or uncooperative tendency, which at base is moral as I use that term.

Tomasello's cartography of primate cognition and human thinking is persuasive in spite of the limited fossil and artifactual evidence. We can, though, reconstruct how the human mind worked in prehistory from individual needs and joint efforts to the collective intentionality of modern human beings. An approach like Tomasello's is important since there is a bridge from the natural sciences to cultural advances in the arts and humanities, critical for any discussion of human development. The core question is how human beings evolved ape-like mental abilities into highly sophisticated social thinking. To emphasize, *social* signifies group behavior in spite of self-interested emotions and sensations. As I have been arguing, individual conduct and group behavior have moral implications.

In *Human Cognition* Tomasello argues that the unique ability of human cognition arises by virtue of a person's development within a rich and stimulating culture. External cultural practices and beliefs then become internalized and work as engines of the cognitive mechanisms. In *Human Thinking* Tomasello takes all of this a step further and argues that the human ability to cooperate, in ways far beyond any behavior in great apes, is what fostered our ability to think. Nevertheless, the ability to cooperate really has much to do with putting aside one's outwardly-expressed selfishness to achieve self needs. Few agree upon any pure altruism and, instead, talk of reciprocal or competitive altruism in cooperation. Some forms of altruism could have evolved to effect conflict amelioration.

In his comparative research between human beings and great apes, Tomasello concludes that a crucial difference lies in the human ability to share intentionality for a common goal. Mind sharing is completed by elements that include advanced capacities for representation, inference, recursive reasoning, and self-appraisal. Culture, he argues, is not just a vehicle to transmit ideas and practices but is one of the main drivers for bridging different minds together in cooperative activities. Like Chase and Donald, this mind sharing has its roots in early people dividing labors in hunting and foraging missions. At the bottom line, what we call thinking is an ability for the individual working with other minds to invent cognitive opportunities, strategies, and answers to complex problems of environment and group living.

In some ways these ideas support my argument that artistic cultural manifestations carry moral import. The cognitive-emotional expression of the individual or the group concerns what is good/bad or right/wrong. What has worth for the individual and how to express or share it? Who approves or disapproves of any activity or creation?

Preliminarily, Tomasello offers a background in which he discusses his ideas in the tapestry of other thinkers, from Hegel and Pierce to Bakhtin. He also discusses Piaget and Wittgenstein. Tomasello sees these thinkers as "social infrastructure theorists" who envision language and culture as a layer that merely *varnishes* humanness without questioning any comparative cognitive continuity with apes. I agree.

Great apes are capable of understanding causality and intention on a limited, personal scale, and the human capacities for advanced culture are not without deep and distant connections to some ape behavior. The differences between ape and human behavior are, of course, pronounced. Human infants are cognitively aware and functioning beings in ways apes are not, before language and culture have set in, proving that there are innate mental modules. The major difference between apes and human beings is that we are able to coordinate and collaborate in a truly collective manner and, therefore, construct great physical objects and institutional enterprises. Apes tend to look out for themselves even if they are functioning in a group.

Does this imply that advanced human morality is Utilitarian, even striving for utopia, with no large dose of self-interest? We will take up that question with Joshua Greene in the next chapter. Nonetheless, higher human cognition, says Tomasello, like recurrent thinking about what others are thinking and a tendency to communicate intentions *for* another person, are built on evolved intuitions. So however mighty we have become in the moral and cultural spheres, our cognitive roots, in spite of mother-infant and pair bodings, are deeply embedded in self-interest. We care, but we really care about ourselves. And that is where my argument begins. Artistic culture is an imperative that springs from our social, symbolic nature. Further, art behavior is a moral behavior for both creator and participant since there is acceptance or denial, approval or challenge.

Human beings, and not animals (though recognition here to Frans de Waal's "Anthropomorphism and Anthropodenial"), can perform objective thinking with representations, argue a point with reasons, engage in social reflection, make inferences, and evaluate mental functions in terms of norms and self-appraisal. These capabilities, according to Tomasello, not unlike Philip Chase's claim, are meshed together by joint mental purposes to accomplish common coordinated goals. Collective achievements are far more advanced than primate joint intentionality geared to individual needs in a competitive but social environment. All I am saying is that artistic culture can at once strive to bring people together or tear groups apart.

Any current difference in spite of mutual prehistoric roots between great apes and human beings is not accidental. Some of our hominin ancestral relations, e.g., Neanderthals, also had capabilities like ours. What we now call culture flowered from a long evolutionary history, detailed in Tomasello's book, spanning "collaborative activity," to "cooperative communication," and culminating in "collective intentionality" (5). Each step of the way is an advance up the ladder of higher order thinking operating alongside moral emotions. For my argument, this activity, communication, and intentionality would carry a moral tinge concerning ethical behavior or not.

Simply because human infants are born with adaptations for cognitive behavior does not imply that collective intentions will occur. Rather, Tomasello asserts, there must be a supportive and nurturing cultural milieu for "cooperative cognition" to develop (6). On the other hand, one's self-interest, while negotiating around this milieu, will inevitably build a different set of neural connections. Natural selection is responsible for cognition, but cognition itself is not necessarily the product of natural selection. The ultimate processes and outcomes of cognition count in natural selection. This does not mean, using Tomasello's example, that for all of its understanding of physical and spatial contexts a spider spinning its web acts cognitively. Instead, there must be very little sureness about one's environment so that, over the course of time, assuming the organism survives, natural selection will empower the organism with the ability to imagine *what if*. My point is that in a social environment, the ability to predict different outcomes can be, with degrees, either altruistic or Machiavellian.

Tomasello offers examples of chimpanzee behavior. As one instance, a chimpanzee sees a tree with ripe bananas, and so internal evaluations and decisions are made about the prospect of getting the food in spite of any real or potential dangers. This scenario becomes social when others are involved who might or might not see the fruit. There is a theory of typing experiences, where an individual stores in memory a schematic action-response which can be called forth in a similar situation, a catalogue of one's life history in feelings and especially images. Some of this schematization can be iconized in a species, as in the example of the chimpanzee and the bananas. For human beings, this picture is more complicated. One's self-interest has tinges of moral behavior, and mental schematics can be built on a series of self-interested behaviors.

With these schemas available, *as if* one is looking at a sketch of one's past, inferential thinking about possible outcomes begins, and the individual, true in chimpanzees, bonobos, and orangutans, essentially diagrams a scenario. This thinking combines evaluating causes, weighing conditions, and predicting results, a cognitive process. Human beings and great apes have much in common, notes Tomasello, from morphology and basic brain construction to emotions, so there is clearly an evolutionary connection, and we can see in certain experiments that apes and human children behave similarly.

Great Ape Cognition

Neurobiologist Kathleen Rita Gibson says the human brain evolved to permit expanded processing. Perceptions, concepts, and motor functions could work simultaneously and extend operations across domains. For instance, while Köhler demonstrates that apes can stack boxes with the purpose of getting food, they do so by chance. YouTube videos of Köhler's apes show how haphazardly they perform. While apes can interact socially with several others, human beings interact in multiple hierarchies at once. Large brains enable us to interconnect more neural networks, vital in terms of intelligence over instinct, so that there are more and better responses to various stimuli.

Cheney and Seyfarth, again with reference to *How Monkeys See the World*, offer insights. Genes and time scale put us much closer to, in this reverse chronology, chimpanzees and bonobos, gorillas, and orangutans. But it is also useful to look at monkeys like baboons, macaques and vervets, from whom our ancestors split much earlier. And of course Frans de Waal would insist that we are perhaps closer to bonobos with our empathetic tendencies. Moral emotions are evident, at any rate, in all the great apes, and in fact one can use accurately the expression cognitive emotions to categorize human beings and apes since emotional responses can be modulated, or not, by thinking.

Deception can be related to cognition. Female chimpanzees suppress sounds when mating with a lower ranking male. Chimpanzees will suppress calls when there is not enough food to share with others. These are examples, say Cheney and Seyfarth, of how nonhuman primates "monitor and predict the mental states of other animals and the consequences of their own behavior on the behavior of others" (192). Does this mean that monkeys and apes can attribute states of mind to others, knowing that another has beliefs and emotions that might differ from one's own? Evidence, according to Cheney and Seyfarth, suggests yes. De Waal insists that great apes are quite aware of how their signals can affect others, a form of mental calculation. Experiments show that chimpanzees can impute mental states to others, as seen in their helping behaviors and especially when they make movements and gestures similar to others who are trying to complete a task or solve a problem, according to Goodall.

Technically, a monkey can understand social relationships, can strategize, can deceive, and can communicate without theory of mind. The ability to hold two different minds simultaneously, one's own and another's, is an essential higher order thinking skill found in human infants. Cooperation and deception imply understanding on some level the mental states of others. Chimpanzees, Cheney and Seyfarth suggest, and other apes are good at learning and reading social intentions because they can attribute purpose to others. Chimpanzees, too, unlike monkeys, will comfort others and care for the sick and disabled because they can read minds. Mental attribution, predicting motives, and theory of mind all have adaptive advantages concerning reproduction, i.e., competition and mate choice.

Monkeys are not as self-aware as chimpanzees and do not seem able to grasp another's motives. Monkeys deceive only to affect another's behavior, not thoughts. Tomasello says apes evolved in an environment where there was competitive foraging for food, so some skills selected for include "spatial navigation," "feature recognition," and miscellaneous abilities associated with cause-and-effect and enumeration (15–16). All apes, Tomasello reminds us, compared with other mammals, are skilled with simple tools and can even use one tool to get or make another. Chimpanzees, in addition to comprehending causality, can infer from effect to cause.

In the social world, cognition in primates evolved in competition scenarios not just for food and other assets but for sexual partners. In *Descartes' Baby*, developmental psychologist Paul Bloom, with a strong reading of the idea of sexual selection says that women are the better mind readers. Women are keenly aware of another's emotions and can make better judgments about another's beliefs. Not coincidentally, men tend to be more physically aggressive, even as children, and are the ones who are more prone later in life to commit murder. By age one a girl makes more eye contact more often than a boy, and the more prenatal testosterone there is the less eye contact, according to Bloom. From a very young age girls are more empathetic than boys, says Bloom. These differences confirm Darwin's theory of male competition in sexual selection and female judgment of mates.

Cognitive functions evolved within a defined group so that individuals and their respective relationships could be differentiated, hierarchies could be established, and coalitions could be formed. In *Chimpanzee Politics*, de Waal says that in terms of coalitions, chimpanzees never act without *calculation*. Coalitions can and do change. Here too, as in the physical world, the advantage selected for is computing outcomes. Apes know that others know and might try to guess *what* others know. Not only does a great ape possess intentions but he knows that others have intentions and most likely have similarly self-interested motives.

Knowledge about knowledge, and especially guessing awareness, is what drives all of our cultural practices and creations and impacts our social and moral behaviors. Tomasello, along with the other forms of social cognition outlined here, concludes that apes indeed think. Note, however, that Biruté Galdikas would say in the case of the male orangutan, because of his solitary nature, there is only the power of his own intelligence to rely on.

Tomasello rehearses all of this research to illustrate that apes are not acting mechanically to stimuli but are making choices in how to act for the best outcome. Other studies, Tomasello notes, especially with chimpanzees, show a variety of cognitive behaviors. For example, they will postpone reward-taking under the assumption that waiting will yield a better result. They will alter previously successful behavior in new circumstances. They will swallow the bitter pill, so to speak, if they know there is a reward forthcoming. They will persevere through difficult situations. These behavioral examples are equivalent to those found in a three year old human child and certainly in many human adults.

Regarding human-like thinking, Tomasello tells us great apes are capable of a number of capacities. First, there is an aptitude to represent previous experience in images and schematics to determine situational pertinence. Second, there is a facility to infer from representations, to make postulations about others. Third, there is the tendency to review one's own behavior in a way that permits a recalibration of knowledge and effort, a cognitively supervisory function. Tomasello is convinced that our common ancestor with the chimpanzee, as well as our ancestry with *Australopithecus*, was "individually intentional and instrumentally rational" before the advent of distinct human culture and so laid the groundwork for advanced thinking (30).

Great apes are aware of their limitations and will self-correct information gathering and behavior to achieve an especially prized goal, in spite of daunting circumstances. This observation falls in line with my argument that human culture is soaked in cognitive emotions that have shades of moral sentiments influencing behavior.

From Self-Interest to Self among Others

The key difference here is that while apes tend to be self-interested with competitive tactics, human beings tend to be helping and caring with cooperative motives. Psychologist Roy Baumeister refers repeatedly to the caring and helping tendencies of the cultural animal, humankind. But care and attention to others, in the ideal world, would have to be pure altruism, which it very often is not. Nevertheless, Tomasello's point is that in the hominin line, the pressure was for a collective sociality that led to sharing intentions for a common goal and thus the rise of thinking. Certainly this was a long evolutionary process. Clearly there are continuities with apes, but to arrive at the complex level of collaborative behavior found in human beings there was a special type of collective behavior selected for, whether early forms of hunting or artistic cultural practices.

Nonetheless, thinking implies individuality, and individualism indicates behavior not necessarily directed to or for a group.

In terms of cooperation and the shifting timeline, Robert Cieri et al. argue that a feminization of craniofacial structure stemming from reduced testosterone or androgen resulted in greater social tolerance from the Middle Pleistocene onward. Our modern behaviors and practices, from sophisticated tools and better resource processing at about 80kya derive from an increasingly non-aggressive temperament. This, then, is a demographic and non-cognitive model for cumulative technological evolution. Larger populations facilitated greater innovation, and social acceptance helped spread advancements.

Complex cooperation is how we differ from apes and other animals. Tomasello points out that there are many species that will act cooperatively, like eusocial insects (e.g., reproductive castes). Biologist Richard Alexander, in *The Biology of Moral Systems*, points out that human beings are not eusocial since we place much value on the individual. Apparently from the ape proclivity for self-interested behavior, only human beings evolved dimensions to identify and accomplish objectives with single-minded synchronization, notes Tomasello. We share an evolutionary continuum with monkeys and apes, but while we grew to adulthood our chimpanzee cousins have remained in mental infancy. The fact is, however, we share a number of tendencies and behaviors, not least of which is acting with self-interest.

Experiments demonstrate that chimpanzees and even bonobos prefer to eat alone, and if there is a dispute about a piece of food, the dominant individual succeeds. Generally speaking, food getting among the great apes is a frenzy of power struggles. Tomasello questions whether or not chimpanzee group hunting of monkeys is really a collaborative effort. Instead, he posits, each chimpanzee is most likely out for himself and considers himself, in advance, as the victor of the chase. On the other hand, there is true collaboration in modern human foragers with their child-care, information sharing, teaching, group decisions, and community organization.

Cooperative behavior probably had its beginnings circa 2mya when the first *Homo* species appear because of competition for food resources from an explosion of other hominids. Generally speaking, and discounting preferences based on rank, mountain gorillas do not compete for food. Silverbacks typically avoid physical confrontation unless defending the group, according to Fossey. Fast forward to the ancestor Tomasello spends most of his time considering, *Homo heidelbergensis*, who appears perhaps as early as 500kya and is an ancestor likely related to Neanderthals and modern humans. But see Antón who questions prevailing views, as well as Matthias Meyer. As even anthropologist Christopher Boehm would agree, *Homo heidelbergensis* is the first species to engage in serious big game hunting, and this large group practice demands collective activity over individual needs. Foraging, too, becomes more than a mere group activity. Individuals are now linked because of common needs, and with interrelated working relationships social pressures increase.

The effect of more complex social relationships, labor divisions, and group efforts is a rise in neural activity and brain organization. In *Chimpanzee Politics* de Waal talks about *triadic awareness* among chimpanzees, the ability to perceive how others relate so as to interpose oneself between them. We see, then, in individual feelings and shared emotions the beginnings of what we now call moral behaviors.

An individual in a social structure might consider, for example, who will be worth working with and who not. Cheater and free-rider detection and suppression begin. The alpha male's power diminishes, and so there is a concern for how one appears and hence, recursive thinking about what others might be thinking. To mark this different behavior, Tomasello looks again to his strength in great ape research. Whereas a three year old human child will share a joint goal with a collaborative effort and portion the reward, this is not necessarily so with chimpanzees. Moreover, a three year old child comprehends division of labor and seems to grasp the notion of obligation to such a degree that if she defaults in the mutual enterprise she will offer an admission of guilt. Chimpanzees, on the other hand, simply gravitate on their own to abundant food sources.

Joint effort implies joint attention, which also means that each actor is attentive to the other's thinking, and both of these modes of attention probably developed together in human evolution to produce perspective taking. As noted earlier, nascent forms of perspective appear as early as one year in a human infant who can see someone else's directed attention, but not until a child is four can she understand that there is a difference between her perspective and someone else's. Tomasello's research reveals that great apes know of another's intentions but don't engage in joint intention. He also says apes are aware of and assist in goals but don't collaborate, so that only human beings have the concept of doing something together. This is where art behavior and artistic culture fit in, since both the creation of and participation in culture involve joint attention and intention guessing.

Since the perspective taking ability manifests itself early in human children, far less pronounced in apes, we can assume that it is an evolved adaptation that helped early human people in collective activities. Accounting for different perspectives is also related to social pressures which gauge how cooperative one is, or not, and helps keep one's own potential anti-social behavior in check through self-assessment. Cultural productions are, in part, replicas of these types of moral norms that strive for social interaction and yet challenge group norms. Although culture implies cooperation, there are always individual self needs and desires at bottom. Moral complexities and conflicts are implied in our cultural practices.

Our thoughts, movements, intentions, and actions pit the needs of self against those of others so that emotions are, in effect, moralized. Morality is a behavior. In turn, for highly creative types, this behavior becomes art culture. For most people, participating in artistic culture with approval/disapproval responses is also a moral behavior.

Social thinking skills are willful control over self-interest and the intention to manipulate others. We know that intentional communication is negligible in apes, who are mostly limited to gestures of direction and appeal. In contrast, Tomasello notes, some experiments show that a human child no more than one year old can point to inform, with a key difference that apes are not able to grasp a relevance inference but the child can. Artistic culture is a means of pointing and inferring, often even a way to deflect or alter attention and behavior.

Gestures and Collective Intentionality

While apes monitor themselves cognitively, we tend to monitor ourselves socially, but social interaction involves cognition. Tomasello says that we estimate whether or not the person we are communicating with understands, which means that we actually simulate to ourselves how the person might respond. The implication is that this simulation is an evolutionary advantage built on top of the earlier concern with social self-image. Even a one year old human infant, experiments reveal according to Tomasello, is capable of guessing, and therefore thinking about, another's point-of-view merely from eye movements. Apes are incapable of comprehending many gestures since they do not grasp intention outside of context, and this failing is related to their inability to engage in highly collective, cooperative activities. Our advanced sharing and caring tendencies do not negate our self-interest but only complicate matters on a neural level, which is why Robert Wright has aptly labelled us *the moral animal*.

Iconic gestures, Tomasello tells us, like raising a hand as if to eat, more than pointing, require both doer and viewer to imagine something else somewhere else, and this imaginative-pretense helped human cognition evolve. Furthermore, Pamela Douglas and Liza Moscovice report that female bonobo gesturing can mitigate social tension, with pantomime movements that share intentionality as pre-linguistic communication.

Iconic gesturing could be ambiguously interpreted, and so the combination of gestures or a gesture linked to a vocalization evolved. Even human infants can use a pointing gesture while simultaneously vocalizing, the two clearly linked by the child for the benefit of the viewer. Here, Tomasello says that what arose in our evolutionary history was *common ground* between gesturing/vocalization and the recipient, a precursor to language necessitating some form of thinking.

Iconic gestures like head, hand, and body movements are symbolic in that they are representations of something else or some other action. Icons make propositions, and therefore engage both doer and recipient in shared, abstract thought, says Tomasello. While apes understand cause and effect, they do not fully comprehend what another is thinking and hence do not engage in the sophisticated collaboration evident only in human activity. The linking of iconic gestures with vocalization, and the mind sharing that outward expressions generated, are what marked our species as symbolic. This also determined for us, eventually, new brain organization.

Thinking about another's thinking leads to one *intending* that another *knows* something, remarks Tomasello. Great apes can only think in terms of themselves. The past is in terms of what is wanted or needed now. Early humans, in having the ability to combine iconic gestures and vocalizations to establish shared communication, were trying through perspective taking to imagine what another might be thinking, so the combination of aspects of cognition led to self-reflection. For the moral and cultural dimensions of our discussion it is crucial to note the emphasis on the presentation and positioning of self in the social sphere. The key point is that presentation and other cognitive elements yield mental representations. Icons, communication, and creative thinking fostered art behavior.

There was in early human cognition a concern for another to understand what one was attempting to communicate, which gives evolutionary rise to rationality. Any concern about what others might be thinking leads to normative behavior. Of course norms imply deviation. Tomasello is not so convinced that any type of so-called Machiavellian scheming prevailed among the *Homo* lineage. If anyone reads the literature on moral behavior, in the next chapter, at some early point a thinking creature realized that he might benefit by taking slight advantage of another. Thinking is not simply representation and communication but strategizing.

Alternatively, Tomasello posits the multi-faceted cooperative, collective attitude distinguished from ape self-interest. But a tendency is only a leaning. One might act cooperatively but have strong, selfish urges. According to Steven Pinker in *The Better Angels of Our Nature*, we are now less violent and aggressive, but that is true only by statistical measure, by data. We measure outcomes. On an emotional and even cognitive level any agent will always consider first what is advantageous only to him, and so there are still many aggressive urges that are latent or deliberately repressed for tactical reasons.

Moving past *Homo heidelbergensis* and the arbitrary 500kya marker, Tomasello discusses *collective intentionality* and how early modern human beings, through the creation of distinct groups, generated shared cultural practices and standards. Importantly, there was collective maintenance and transmission of culture over time. While some can argue that chimpanzees and orangutans have culture, Tomasello says that these behaviors are not taught cooperatively as in human culture but appear "exploitive" (82). There is haphazard and inadvertent copying. However, a team of primatologists that included Carel van Schaik and Biruté Galdikas determined that great ape culture, from tool use and feeding to social signaling, has persisted for 14 million years.

The bottom line is that apes are social but human beings are cooperative and, therefore, there is full-fledged human culture. Nonetheless, sociality implies self-interest in a group context and what could loosely be labelled as moral emotions and moral behavior. While we don't think of chimpanzees as moral, they do evince a sense of what works for them as individuals in the group. See, e.g., de Waal, *Chimpanzee Politics*, which is filled with illustrations of individual maneuvering, for good or bad, among the group. Although chimpanzees and bonobos do not have culture like ours, their self-interested but outward-directed conduct is at root in our cultural behaviors.

Our advanced culture is an extension of primate behavior that can work for the group but certainly does not eliminate individual needs and desires.

Communication as Selective Force on the Brain

Tomasello discusses his notion of the ratchet effect, which states that while there is fidelity in cultural transmission, any individual might devise an improvement on a practice, which, over time, can itself be improved, and so forth for many generations. The fine-tuning of ideas, values, beliefs, and practices leads to more closely-knit groups and the splintering of groups, evident even to this day across nations. So what is moral infancy among great apes becomes moral complexity for our species. Has anyone tried the famous runaway trolley scenario on chimpanzees? In the discussion of Greene, to follow, we will take up the trolley scenario.

The shift is from what Tomasello earlier identified in human prehistory, circa 400kya, as the recognition of the second person (me-you) to now, at 200kya with AMHs, a group identity. Common ground is shared among various individuals to the extent that one can be marginalized if not cooperative. Tomasello suggests that group formation and cultural adhesion will play an important role, eventually, in directing the successors to early modern human beings toward an objective point-of-view. There is a move from the one-to-one mode of early people to the normative group thinking that fosters emotions like guilt and shame. Christopher Boehm, too, discusses this movement to shaming as a cultural norm. While this is true, there is always an individual and his or her needs, desires, and motives competing with the group. To consider us as a tight flock of birds or as a synchronized school of fish is inaccurate.

Some cultural practices can turn into institutions, with a sub-group being formed to make certain decisions about food sources or defense. There is not much further to go, then, to a band within a group that might have been entrusted to implement more powerful tasks, like making judgments about behaviors or resolving disputes. Eventually, the logical consequence here is the literal and figurative creation of a status-symbol leader, concludes Tomasello. Through these emerging groups linguistic communication began to become dependent on cultural commonality. There is an understanding that in communication one would employ the common verbal cues of the group to make meaning unequivocal. Over time, any communication gestures, in whatever form, would become reduced and synthesized into particular languages and hence fence off groups.

The point, then, is that complex, symbolic communication gave rise to reasoning abilities and the capacity for one to convince another of his position on an issue or decision. Key here is that reasoning is collaborative back-and-forth communication, which would put more stress and selective force on the brain. One who argues a point will anticipate the opposition and so have an internal conversation about positions and objections in order to succeed and obtain an advantage. Behind this behavior is the idea that we moved from unknown factors that had to be reasoned out in accord with social standards, to shared norms and linguistic styles which enable one imaginatively to propose a workable theorem about another's point of view.

Thus flowers the feeling of individuality. A single thinker can step aside from a thought to evaluate it through reason in relation to group culture and subsequently, suggests Tomasello, further recalibrate her reasoning in light of any new ideas. The important evolutionary cognitive and linguistic development of objectivity commits one to reach back over her thoughts in collaborative communication. Objectivity does not rule out individual gain, though, even if minimized or deferred. By its nature, social culture restricts practices. And yet artistic culture can offer any number of perspectives.

Tomasello's basic thesis is that because of environmental pressures, like competition from other groups, climate variation, and changing supplies of food resources, early human people increased group size and began a movement to more collaboration. Human thinking is equivalent to cooperation, similar to what we saw in Chase. Tomasello's emphasis is on the evolution of culture via cooperation, including early forms of sociality, division of labor, child care, pair bonding, and group foraging, all of which by 50kya evolved into highly complex joint operations on an increasingly larger scale.

Without discounting any of this, the view here is that human cognition includes individuals who will cut against, even if subtly, any group norm or cooperation and forge a different perspective. I am not necessarily talking about cheaters, who are culled out. Rather, moral culture means being able to see, consider, and possibly embrace another viewpoint or representation. In-group competition is not baneful but engenders social behavior like the perspective taking of another's position.

With great apes any collaboration or communication is essentially for self-survival among aggressive competitors. While male orangutans will fiercely guard their territories, and while male mountain gorillas will mortally fight for females, chimpanzees do not necessarily fight with strength but with mental control to achieve cooperation. De Waal tells us that while chimpanzees engage in dominance struggles, where dominance can be either for physical space or influence, alliances and reconciliations are equally important since one never knows who, in the future, he will need. Leadership and equilibrium are cautiously valued in the group. Our species is endlessly engaged in mental gaming with each other and with other groups.

Through a common goal came shared attention while maintaining one's own point-of-view. A mutual communication style thus arose with gestures and pointing which forced inference to bloom on a joint intention far beyond any similar capacity in apes. This type of recursive thinking, the implications of an implication, fosters one to become not only self-aware but also aware of what he represents to another mind, whether his motive is good or bad.

There evolved a set of *mores* and practices so that different, unknown members of the same group could work together while distinguishing themselves from members of other groups, hence the rise of group personality. Early forms of shared intentions most likely existed in Africa prior to Neanderthals. Later, intentions laid the groundwork for cultural differences with the spread of groups out of Africa. Even the cognition of *Homo sapiens* could vary depending on local needs and practices, but all distinct groups evolved culture in a cumulative manner that was tied to cognitive development.

Was Darwin Wrong?

Was Darwin wrong to draw continuities between apes and human beings? No. The preponderance of the evidence only hinted at in this book proves so. Nonetheless, Derek Penn, a former stock trader turned cognitive scientist, along with others say Charles Darwin was incorrect about any ape and human continuities. Their paper, "Darwin's Mistake" emphasizes the "profound discontinuity" between the ape and human mind.

While the authors attempt to discredit a number of recent researchers in this area, they do not cite either Wolfgang Köhler or Robert Yerkes whose pioneering work in the early part of the twentieth century was done on chimpanzees, monkeys, and one orangutan. And they were not the very first. After thousands of experiments and tests, Köhler, for instance, became convinced that chimpanzees think abstractly. Yerkes goes further and says that apes not only think with ideas but operate rationally.

Need we repeat how genetically close we are to chimpanzees? Köhler and Yerkes, of course, would not have had the sophisticated computational models of Penn et al. In fact, some commentators to Penn suggest their model of discontinuity is too complex and only proves itself without proving anything. Of course we are smarter than apes.

Louise Barrett says Penn makes the error of assuming any cognitive leap was only internal and not external. Derek Bickerton says Penn's model does not explain the evolution of any discontinuity. Most telling from a host of negative reactions to Penn, Thomas Suddenorf notes the obvious. He says that Penn et al. see a discontinuity because they ignore, as do many others, the various prehistoric species related to us that have gone extinct. For instance, again I reference Antón, Berger, and Dirks. And Edward Wasserman prepares us for the next section, especially pointing to skeptics of ape and human continuities, by demonstrating that Penn resuscitates an intelligent design argument for the uniqueness of our species (as if we evolved from human beings).

There are many research articles that demonstrate cultural or quasi-cultural actions in chimpanzees and bonobos, indicating their proximity to us. Edwin van Leeuwen et al. discuss how one chimpanzee put a strand of grass in her ear, and this was copied and spread within the group. Jan van Hooff and Bas Lukkenaar explain how a chimpanzee deliberately plans an action with a tool to strike a moving object. In two articles Kathelijne Koops et al. conclude that chimpanzees, like us, are intrinsically inclined to manipulate objects. Moreover, there are cultural differences between neighboring chimpanzee groups in how ant-dipping tools are selected and used.

Zanna Clay et al. study the functional flexibility in wild bonobos and conclude that a variety of peep sounds clearly express psychological and affective states, a precursor to language. Anticipating my discussion in the next chapters, Robert Latzman et al. demonstrate that, as in human beings, there is a relationship between the five factor model of personality (Openness, Conscientiousness, Extraversion, Agreeableness, Neuroticism) and brain neurobiology in chimpanzees. This means that there is an evolutionary basis in our moral, cultural, and artistic behavioral dispositions. And Daniel Povinelli and Jennifer Vonk go as far as suggesting that the chimpanzee mind is questionably human in how it forms mental concepts of another's behavior.

These examples are the tip of the iceberg in primate research but show how great ape behaviors play into our own cultural and moral prehistory and development. What I call here artistic culture has deep origins in non-human nature.

CHAPTER 2

Biology and Morality as Interrelated

In *Braintrust*, neurophilosopher Patricia Churchland emphasizes how the brain chemical oxytocin is one key part of our moral system. This ancient neuromodulator, originally involved in reproduction, is widespread in vertebrates. Only in mammals is oxytocin implicated in infant care, later extended to prosocial emotions. Churchland says that oxytocin activates in the mother's system under any stress brought upon her infant. Importantly, neural changes would have been in many species and would have undergone multiple evolutions before attaining the greater caring and helping tendencies we now see.

In a profound argument against selfishness and for cooperation and empathy, Frans de Waal in *The Age of Empathy* says that parenting extends to about 200mya. The desire for children might stem from innate caring tendencies. In other books, like *Good Natured* and *Primates and Philosophers*, de Waal says that our morals have precedents in the cognition and sentiments of great apes and like them in our interdependent social nature. Moreover, apes, both in captivity and in the wild, do not have to be closely related to work together, as long as assistance can be reciprocated. The ability to guess another's intentions probably arose from cooperation, not from competition, and might be implicated in how we have sclera (the whites of eyes) over apes. Not only do we read faces, but we read eyes to judge aims.

What we call morality has, as David Hume saw, a natural foundation and is not exclusively rational. Reason only modulates emotions. Churchland correctly says that morality may be rooted in biological "attachment" (23). As parents began to accept each other as mates, Churchland goes on, brain chemicals like oxytocin and vasopressin among *Homo erectus* to *Homo heidelbergensis* and especially in *Homo sapiens* solidified pair bonding.

No doubt pair bonding arose so that a male, who would offer protection to a female and some parental care, was granted exclusive mating privileges with her. There are costs to both males, e.g. energy loss in attracting and protecting a mate, and for females, e.g. pregnancy and infant care, so that cooperative behaviors evolved to mitigate high costs. Over time, people began to realize that cooperation is the best strategy in terms of getting and using resources. Basic confidence between parents and the benefits bestowed to offspring eventually spread to other kin, relatives, and group members. Here we see the seeds of cultural behavior.

 | DOI 10.1163/9789004319486_004

Although we are biologically evolved to be pro-social and involve ourselves in the care and aid of others, skeptics discount any evolution of morality and adhere, instead, to abstract and metaphysical rules.

Naysayers Concerning the Evolution of Morality

Here's the question posed by philosopher Hilary Putnam: How can evolutionary biology help us understand ethical judgments? His reply, while acknowledging moral emotions, is "Not very much." This response mistakenly looks for a unified answer. Evolutionary psychologists Kelly Asao and David Buss have answered the question by positing cumulative adaptations that include moral judgment, moral influence, and moral conscience. Moral judgments are not human-created but outcomes in a long chain of primate, cognitive adaptations.

Natural selection and other pressures provide the hardware and behavioral motivations for approval/disapproval emotions which harmonize with artistic culture. Sharon Street raises questions about the compatibility of human values with Darwinian science. While she acknowledges the impact of evolutionary forces on how and what we evaluate, those forces in her view did not affect the content of human values. However we look at it, there are philosophical implications in any discussion about Darwinian evolution. What follows is more analysis than argument. I'll return to Street at the end of this chapter.

Psychologist C. Daniel Batson asks what motivates care and concern expressed not only for family and friends but for others and even animals one has never seen. Do we care for others only out of self-interest? While Batson favors social constructivism he says any social learning does not answer the question of altruism. Acting *instrumentally* for others we nevertheless consider our own welfare. He admits that his view, contrary to current evolutionary thinking, lays emphasis on social learning and not on inherent mental faculties or modules.

Batson sees *principles* that make us interested in the well-being of others, but these principles are not necessarily *moral*, so that altruism is a motivation and not a helping behavior. Along with Sober and Wilson, Batson maintains that there is evolutionary altruism (increases another's fitness) and then there is psychological altruism (ultimately increases another's welfare). There is no need for evolutionary altruism to have psychological altruism. His point is that while one might be altruistically motivated, the end result is egoistic. Batson, on the one hand, offers a skeptical viewpoint of moral behavior, and, on the other hand, offers self-interested biological egoism which is the anxiety of some philosophers like Hilary Putnam and Susan Neiman.

In a paper on *moral tension*, biologists David Lahti and Bret Weinstein argue that discussions about altruism and selfishness create a false dichotomy and ignore the larger continuum between provisioning for the individual or the group. Following Alexander, Lahti and Weinstein focus less on any evolution of morality as a common good or as a war against cheaters. Instead, an individual's competition in or commitment to the group depends on the group's stability. In terms of my argument, this line of thinking fits nicely, since artistic culture can be an expression of cooperation (status quo) or protest (reappraisal).

Batson questions any type of moral motive, i.e. moral action from reason, emotion, or intuition as suggested by Antonio Damasio, Jonathan Haidt, and Marc Hauser. So-called moral people can act immorally, as demonstrated by the obedience experiments of Stanley Milgram and the conformity experiments of Philip Zimbardo. As well, says Batson, there is "moral exclusion" (selectivity), "moral oversight" (self-serving), "moral rationalization" (saying one does a good deed but the same act by another is deemed as bad), "moral disengagement" (a sort-of turning off), "moral hypocrisy" (appearance over action) (51–52).

But this analysis, by Batson's own conclusion, restricts moral behavior to what is learned and not to any genetic basis whatsoever, the latter of which he labels as speculation. As biologist Mary Jane West-Eberhard has argued, through individual (phenotypic) plasticity, itself an adaptive response to the environment, diversity and new traits for adaptation arose, where behavior is particularly malleable. So if in any environment, whether physical, internal, or among conspecifics (those of the same species), a behavioral response routinely enhances fitness, then any underlying genes will be selected for.

Behavior can evolve and even influence morphology, West-Eberhard says. Flexibility in the individual or phenotype can originate social structures, the view taken here, particularly with the evolution of artistic culture. Adaptations are "reliably developing," says Buss (39), which explains why artistic culture went from simple to complex, from individual to social expression.

Jeffrey Schloss sees an evolutionary basis for various aspects of morality, like "affective dispositions" or "cognitive capacities." But contrary to E.O. Wilson and Frans de Waal he does not see morality as a genetic enhancing mechanism but as a byproduct of gene/culture co-evolution. While Schloss deserves some credit for hard-pressing evolutionary explanations of morality, he slides into skepticism.

There are "philosophical issues" that science is incompetent to mount, says Schloss, taking particular aim at de Waal. Schloss, a biologist, says any presumed evolutionary explanations of morality are by nature insufficient when they try to pinpoint a unity, for in reality what we call morality is really an ambiguous range of sentiments and intuitions acting in concert. Schloss says that while evolutionists observe or discuss empirical behaviors, they interpret those behaviors incorrectly. Chimpanzees and bonobos are not the only species who exhibit care for kin. Schloss claims that this fact only blurs what has been called proto-morality.

The sense one gets from reading Schloss and other skeptics is that while they posture an evolutionary stance they exclude the human mind and surely moral behavior from evolution (whether by natural or sexual selection). Part of their skepticism resides in an unfounded fear that evolution will take away free will and determine behavior.

Schloss further says that even granting primates empathy does not make them moral, since empathy is strictly speaking a "morally neutral capacity" (86). True, one can feel into another's emotions solely for personal gain. A caring or helping stance might even be the unintended result of some other goal, like coalition forming. Besides, Schloss says, if we follow de Waal's thinking logically, many biological processes could be proto-moral, including minute brain chemicals interacting. There is much ambiguity concerning where to begin. Certainly, for if atoms and cells did not cooperate, then of course there would be no matter. And as we know: no matter, no mind.

Unfortunately, none of Schloss helps one understand moral sentiments from a Darwinian perspective, worth mentioning since he, and others I've dubbed naysayers are included in a book called *Understanding Moral Sentiments: Darwinian Perspectives*. Schloss goes on to show how even the term altruism is riddled with ambiguity depending on which discipline uses it and how. Psychology and philosophy speak of altruism as a motivation (benefit) while some evolutionists speak of it in terms of a consequence (cost). Schloss says any explanation of fitness enhancement does not necessarily explain fitness-subversion, so that while biological altruism might have selective force, it does not completely explain what we see as human morality.

What's more, Schloss sees culture as completely Lamarckian and not Darwinian so that cultural change is not via natural selection of small and inherited variations that enhance fitness in reproduction and survival. While at this point many aspects of what we call culture are probably Lamarckian, the basic units of culture – visual representation, music, narrative – are not, since they would have been eliminated as useless via natural selection. Schloss says that many explanations of cultural transmission are after the fact and like Batson leans more toward "socially mediated norms…" (93). But is there not some evolutionary basis for this transmission? While one might not care for a particular television drama, one like many others does care to witness actors perform imaginary lives, and it makes less sense to dismiss this fundamental human desire as an epiphenomenon.

The five moral sentiments of Haidt so carefully and clearly outlined by Pinker in the same book (to be discussed) are deemed unclear. As for Boehm's notion of moral aggression and conscience (to be discussed), Schloss says any apparently cooperative social-building sentiments could easily have been *underwritten* by the urge for dominance. Schloss thinks any moral issue on the table for discussion requires philosophy "to assess moral sentiments" and the intuitions they attend and biology to explain how sentiments *form* and become transferred in a culturally flexible manner (97). His essay unravels with a critique of Richard Alexander and then mini-sections laced with religious references, which leave the reader with little constructive science about evolution and human culture.

Theologian Stephen Pope makes an unlikely pairing of Darwin and Aquinas in writing about moral decision making. While Pope acknowledges that moral traits are evolved from *social instincts*, he places too much faith in any natural aspect of moral sense in leaning to Adam Smith over David Hume. Approval/disapproval need not be rational but only a physical sensation. It was the Earl of Shaftesbury in discussing art and aesthetics who first forged the notion of the moral sense in British philosophy. Voland and Grammer remind us that in the Western tradition what is beautiful has often been associated with what is moral. Pope does, however, acknowledge that Darwin recognizes the predominance of our emotions and feelings over rationality.

After a discussion of Haidt, Pope is critical of psychologist Marc Hauser's supposed presentation of religion merely as a collection of laws. Pope sees Hauser's morality as judgments and not as virtues. But his analysis is not, as it purports to be, a Darwinian perspective. While there is, over Haidt, a cognitive flavor to Hauser, it is perhaps overstated. Pope moves squarely to Aquinas who based morality in the love of God as well as in others, and so love becomes a rule and not an emotion "to will and act for the good of every person and for the larger community" (158). To invoke God is not an explanation but an excuse to avoid any other rational, factual explanations. God is not present in Darwin's world picture, which is not, considering his Victorian frame of mind and devout wife, to say God is non-existent at all.

So it is difficult to read about moral sentiments from a Darwinian perspective with abstractions like virtue and God. Granted Aquinas was steeped in Aristotle, but so is Martha Nussbaum, and while Aquinas' emphasis, like Aristotle's, on character is compelling, any discussion without developmental psychology, genetics, or evolutionary chronology (i.e., the hominid/hominin lines) and artifactual references is a serious lacking.

Contrary to Haidt, Aquinas does not deny body or emotions but sees emotions as *disrupting* decision making. Earlier we noted how current neuroscientists view reason as guided by emotions. Ultimately, Pope is critical of Hauser and Haidt, saying that their emphasis on emotions blinds them to "the proper place of rational deliberation in the good life" (162). Here is a tacit denial that our reasoning capability itself is an evolved function which lies close to and in fact is embedded in our emotional life. Pope's point about ethical teaching and *the ought* is well taken, though for an essay like his that postures Darwinian thought, yet filled with abstractions, he ends on a note that dismisses evolutionists as reductive.

Although the title of an essay by theologian Timothy Jackson contains the words genes, anthropology, and biology, his view is that any evolutionary account of morality is reductive. Pope, Jackson, Putnam, and Neiman, who in some way say that any biological explanation of morality is reductive, do not offer an empirical alternative. Moral explanations need not be taken on faith alone. Part of this dilemma, to be addressed by David Lahti at the end of this chapter, is not only in the silos of disciplines but in the choice and definition of key terminology. The reader sees these disciplines talking, but one is not sure if they are all listening to or even reading each other.

The thrust of Jackson is to bash Richard Dawkins, first *The Selfish Gene* and later *The Blind Watchmaker*. Jackson says that Dawkins is unscientific and he prefers, instead, the ethics of Mark Twain. Specifically, Jackson says that he wants to *reconcile* "Christian charity and biological evolution" (170) but spends his entire essay in attack mode. Are only Christians charitable, neither one of those words used by biologists? Jackson says evolutionary psychology is reductive and needs to be "revealed and rebutted" (170).

Jackson mistakenly, in his misreading of *Selfish Gene*, says science absolves a person from personal responsibility, revealing his ignorance or neglect of, to name only one, neuroscientist Michael Gazzaniga's *The Ethical Brain*. If, as it seems apparent, Jackson wants to argue for Mark Twain's literature as ethical, as Nussbaum has elsewhere more thoroughly and persuasively done with Henry James, why is his writing about morality from an evolutionary perspective? The essential conundrum for Jackson is in his inability to separate DNA and moral behavior, the latter of which he insists on calling ethics (obligations and prohibitions), a theological and not a biological term. While he claims that biology represented by Dawkins has reduced personhood to an epiphenomenon, nothing could be otherwise.

You, reader, are only your genes and no one else's, and that is your distinct personhood, and the burden of your responsibility is how your person will act. The challenge is to use one's multi-lobed brain and one's environment, education, and circumstances in the most responsible way. Genes respond to environment, but at the same time one can, based on genetic disposition, gravitate toward a certain environment. What made Christopher Boehm an anthropologist and Timothy Jackson a theologian? Genes only? Circumstance only? Clearly it was a matter of both working in complex ways, and to dismiss or minimize one's genes in any discussion of personality is just misguided. See, as only one example, the Rimfeld et al. paper on how genes influence academic ability or Polderman on the heritability of traits. Jackson and some others mentioned here seem to believe that one's biology eliminates personhood rather than marveling at how nature, to propagate our species, has in fact guaranteed that we are biologically varied. In his book *Evolutionary Psychology*, David Buss takes pains to show how a major misconception of most people is that, as Jackson suggests, genes determine behavior. Genes are influencers, as is environment, at about a 50/50 split.

Yet Jackson insists that Dawkins (he alone?) "effectively denies personality to human beings…" (172) as if we are all cloned automatons. He says that Dawkins' assertion that the basic unit of selection is the gene is *reductionist* and that "the person is reduced to the gene" (176). Jackson fails to note that no one gene is responsible for any single behavior and that genes act in many complex ways, especially with the many genes and many neural connections in the brain. And he ignores gene activity modifiers from the external environment. Rather, Jackson's language is reductionist. Each human being has approximately 24,000 genes, so how is that reductive? Each brain has about 85 billion neurons and many trillions of synaptic connections. Because of what we decide to see, hear, think, read, etc. we create the synaptic connections. Reductive?

Jackson seems overly concerned about *randomness* as if there were no variation, competition, and inheritance in natural selection. Random is neither bad nor good. In the first edition of *Origin of Species* I am not aware Darwin even uses the word random. Variation, on the other hand, facilitates the spread of genes. Granted there is no design feature in natural selection, and it tends to cobble one advantage onto another, but that is not quite random either. Speaking of the multitude of genes, one ought to mention so-called junk DNA, not yet coded but surely responsible for contributing to some of our traits that distinguish us from our very close genetic relatives, chimpanzees.

Hilary Putnam is not much more helpful than Jackson, though also purporting to write about moral sentiments from a Darwinian perspective. Like Jackson, Putman presents a skeptical view of evolutionary explanations of moral behavior, so unfortunately another non-Darwinian perspective. Interestingly, one of the greatest skeptics in moral philosophy is David Hume, admired greatly by the so-called pessimist Schopenhauer, and often cited by many evolutionary thinkers including the neuro-philosopher Patricia Churchland.

Like Jackson, Putnam fails to see morality as a behavior and clings instead to philosophical abstractions like ethics, a far more complex idea than what Shaftesbury had in mind when talking about a moral sense. To his credit, Putnam admits that evolutionary theory could have something to say regarding "spontaneous sympathy" and "altruistic behaviors" (203). The problem for the writers mentioned in this preliminary note seems to be that they make some mistaken equation between what we call morality or on a higher plane ethics and evolutionary adaptive processes. Or they insist there is no physical or psychological process in moral behavior.

This is why in my Introduction I have tried to be careful in how I define and use the word moral. What we call moral is not an abstraction but a physical and psychological sensation of approval/disapproval. From the kernel of kin sentiments, and from altruism, feelings, emotions, and reciprocity, what we call morality flows. In de Waal's *Primates and Philosophers*, one of the commentators, Christine Korsgaard, drawing from Nietzsche and Freud, says that human beings have broken from nature. That is untrue as demonstrated from Chapter 1 of this book, though a predominant stance of naysayers.

These particular philosophers seem to look back for explanations, but they look to Aquinas and the Greeks and stay there rather than looking at human prehistory. As Darwin himself would insist, like the gradual sedimentation of layers in the earth, no evolutionary account points to full-fledged morality but only to the many contributions, behaviorally and genetically, to what for human beings is now called moral behavior. The only illustration that appears in *Origin of Species* is a branching diagram, in the chapter on natural selection, which highlights the fitful starts and sideway dead-ends of a long, natural process for many species. That chart for speciation could also be used, expanded tremendously, to illustrate the evolutionary branching of artistic and other human cultures. Where any cooperative behavior reaches a different result in an ape it continues to develop in us.

Putnam says any biological explanations of ethics (his word), especially with reference to groups or apes, are inadequate. Evolutionary theory is in the picture but reduced to a corner of the boxing ring. Who then, are the main fighters? We are not really told. In another turn, Putman then says sympathy and altruism are "among the *preconditions*" for ethics (203). Agreed, though ethics might not be the best choice of words, as in this book I keep reminding readers of that slippery word *moral* when used in terms of primates. Like Jackson, Putnam discusses virtue and ethics and Greek thought while getting in quick jabs at Hauser. Coming back to ethics as a cultural product, he admits the process is Lamarckian and not Darwinian.

Culture is a loaded word that includes many variations of beliefs, values, and practices. Culture is in part a biological phenomenon since gene frequency can and has changed based on an organism's ability to successfully reproduce and survive. Whereas Darwin speaks of sexual selection we might speak of, and indeed some do, culture. Some might say social selection. Human needs and desires have, over a very long course of time, constructed culture, from simple stone tools, body decoration, jewelry, carvings, and paintings to the more sophisticated forms of culture today. The human need for cultural artifacts and practices became inborn. Neural-genetic plasticity works on an individual level so that any early formation of a cultural artifact later became improved and spread. But none of this denies that our moral sensations are physical and part of nature.

Susan Neiman, claiming to write about moral sentiments from an evolutionary perspective, begins her analysis with a literary example from Rousseau and never waivers far from that. Like Putnam and the others mentioned here, who supposedly posture a Darwinian perspective, Neiman is clearly a Darwinian skeptic. She claims that evolutionary psychologists do not focus "on the human shape of erotic desire" (213). One is not sure what that means and whether she includes, to name but one, psychologist Geoffrey Miller in her equation. Granted, evolutionary psychologists look at a landscape that often excludes the individual, but some have devoted lots of research effort, in the field and theoretically, to Darwin's notion of sexual selection, like David Buss who has done comprehensive cross-cultural studies.

Like Jackson, Neiman does not put forth any Darwinian perspective and uses Rousseau and Hobbes to attack as reductive anything evolutionary. She denies that any Darwinian account can explain the origin of moral behavior, in spite of what, we shall see, Boehm and Krebs have to say. A good critical examination of the work of Miller comes from Kathryn Coe in *The Ancestress Hypothesis*, but Coe counters Miller's science with science. Neiman, Jackson, and Putnam discount origins, offer no alternative, and merely attack evolutionary reconstructions as "subjunctive" (216).

Astounding it seems that a philosopher, no doubt never using the subjunctive tense, could level such a linguistic charge at all of the paleontologists, anthropologists, and other scientists who over the years since Darwin have assembled a mountain and array of fossil and artifactual data which is in a very active tense.

Recalling twentieth century behaviorists who turned away from Darwin, Neiman insists that one is morally trained and that there is no biology in morality. No developmental psychologist or neuroscientist would discount the important influences of parents, peers, environment, circumstance, and education on the moral education of children. But language like Neiman's only hefts an iron wedge between philosophy and science. Likewise, philosopher Sophie Berman accuses science of determinism and posits different scientific and metaphysical *planes*, where morality belongs only to the latter. E.O. Wilson's dream of consilience is shattered by anthropocentrism and anthropodenial.

Foundations of Human Morality

Primatologists Cheney and Seyfarth, in *How Monkeys See the World*, shed light on how the social behaviors of monkeys and apes correspond to our own. Evidence from studying the behavior of primates in captivity or in the wild exhibits their caring capability toward others in need, bolstering the premise that "moral sentiments are rooted in human ancestry" (Allchin 595). In 1963 Jules Masserman found that macaques would themselves endure prolonged hunger rather than submit a fellow macaque to electric shock just to get food. Claudia Rudolf von Rohr et al. show that bystander chimpanzees strongly react to seeing acts of harm against infants, pointing to social norms.

Cheney and Seyfarth demonstrate, with vervets as well as other primates, kin bonding via grooming and alliance generation from care received by the mother. For Robin Dunbar, our species replaced the one-on-one grooming with language or laughter, which can be informational and bonding. The increase in and demand for sociability among our primate and hominin ancestors led to alliances in larger groups and then a larger brain. For us, according to Dunbar, gossip has an evolutionary history. Language rather than grooming permitted us to keep track of others and what they might be saying about us. What we call moral behavior is unquestionably rooted in nature.

Cheney and Seyfarth go on to say that feeding, grooming, and distress support from the mother contribute, especially for female infants, to later cooperative social behaviors. Kin recognition for most primates is based on the intimate ties offspring develop with the mother, not the father. However, Fossey is strident to emphasize the cohesiveness of the family unit in mountain gorillas and insists that a silverback, who can weigh over three hundred fifty pounds, is nevertheless very gentle with his offspring.

Infants learn their social rank by virtue of the mother, and this status can determine not only who plays with the monkey but how and how often, so that from infancy some hierarchy-based primates learn about their dominance position and potential. With male monkeys, Cheney and Seyfarth say, rank can fluctuate periodically and coalitions can help those in low social stations. Rank is crucial since it entitles one to the best resources and mates, and so hierarchy often shifts because of competition for alliance and grooming. Of course a silverback gorilla does not need to groom anyone by virtue of his high social standing.

While Darwin in *Origin of Species* sees individuals acting in their own self-interest, there are instances, as even Darwin knew, of social helping behaviors and reciprocity. Chimpanzees, for example, do not necessarily exchange goods but favors. Kin selection as per biologist William D. Hamilton provides a clue as does Robert Trivers' notion of reciprocal altruism between non-kin. Natural selection can favor forms of selective altruism where there is some return or expectation. Since reciprocal altruism would rely on memory and future recognition, there is some cognitive mechanism involved, especially since there is a cost-benefit calculation to be made as well. As Cheney and Seyfarth and others would say, there is an "ability to detect cheaters" (43). On a related note, close bonds between a male and a female baboon or macaque can persist for years in that the male provides protection while the female provides reproductive opportunities.

In terms of sociality, while not all primates are territorial, e.g., baboons and macaques, they nonetheless can clearly distinguish members of their own group from nonmembers and hence the competition for status. Primates, therefore, do observe the interactions and relationships of others and attend to ranks. Cheney and Seyfarth say that chimpanzees, baboons, and perhaps additional monkeys notice others in groups since they understand how kin and non-kin relations work and so "make judgments" (61). With female vervets, any decision to render aid is based not only on kin status but whether or not any help or resource has been recently proffered.

Behaviors like grooming can over time "and incrementally" help establish and maintain a relationship similar to conversation and polite and acceptable physical contact among human beings, note Cheney and Seyfarth (69). Chimpanzees, compared with vervets, will retaliate with vengeance against those who form alliances against them in the dominance hierarchy. Assessments are based relative to kinship, dominance, and rank. The ability to make these calculations indicates a cognitive sorting and labeling capacity, grounded in age and sex, which reveals how primates evaluate and predict. Some of our so-called moral behaviors are remarkably similar.

While there are many significant topics in evolutionary studies, the question about the foundation of human morality is paramount. Moral emotions, some of which are evident in other primates, and moral reasoning, which we alone seem to have, affect nearly every aspect of our lives. Because of the deep history of moral emotions, my claim is that artistic behaviors and art culture, both in creation and participation, are not immune from pervasive moral sentiments.

While the contributions of Batson and Schloss might force us to look harder at the question of evolution and human culture, we can see them as outliers. Psychologist Dennis Krebs, in *The Origins of Morality*, offers a careful genealogy, informed perspective, and comprehensive précis of the most important of all human themes, how we came to be intricate and at times contradictory moral creatures.

Krebs studied under Lawrence Kohlberg and worked with Robert Trivers. Kohlberg shadows *The Origins of Morality*, for Krebs says that while he began looking for ways to align Kohlberg's stages of moral development with current evolutionary thinking, he had to abandon Kohlberg's paradigm altogether. In reference to reciprocal altruism, Krebs thinks the correct or more precise terminology should be "the evolution of reciprocity," since altruism is as slippery a term as Dawkins' use of the word selfish (8). Often overlooked is that reciprocity implies returning a harm as well as a benefit. Krebs says a problem with Kohlberg's approach is that "people's conceptions of morality do not necessarily get better as they develop…" so there is no real equivalency between high intelligence and morality (26).

One could argue, as psychologist Nicholas Humphrey has, that the biological function of the intellect is essentially Machiavellian or cunning. Dunbar says the deceptive aspect of Machiavellian intelligence fostered mental representations that contributed eventually to symbolic culture since the mind at once held both false and true ideas.

Mark Flinn et al. say that an autocatalytic process of social competition and coalition forming accounts for the rise of human intelligence in its striving for ecological dominance. Human beings became a key source of self-induced selective pressure on each other. We see this pressure evident in our brain evolution, from spindle cells for social intuition, to widely networked modules in the prefrontal cortex, to scenario simulation in the anterior cingulate cortex. Our super intelligence accounts for how we socially finesse moral emotions.

Our self-interest is not necessarily for malicious intent but more to serving ourselves since we can get what is advantageous by helping and caring. Our psychological makeup requires that we try to see other minds as reflections of ours and through introspection make attributions of needs and desires. Nonetheless, this does not rule out the possibility of near pure altruism, such as anonymously risking one's life to save another. As Humphrey suggests in *Consciousness Regained*, psychological skills have been equally important in cooperation as well as in competition. Even among chimpanzees their complicated social situations imply consciousness and mental comprehension for survival and reproduction. Another animal might have intentions that are similar or different, and conscious introspection and theory of mind help one influence the behavior of another.

Krebs points out that human morality is not about the fact of achieving survival and reproduction, which can be entirely selfish and apparently bad, but *how* one achieves these. The question becomes, what is the cost put on another. Unhealthy, dangerous competition, which is not mutual in any way, creates tensions leading to punishment by others and ultimately destroys any chance for coalition. We have evolved other means to place costs on others to benefit ourselves. Hard as it might seem, artistic cultures are part of this equation, though costs and benefits might be difficult to measure.

While any bleak assessment is based on evidence in nature, at the same time any so-called social contract is itself an innate sensibility of fairness and quid pro quo, also roughly apparent in nature. Hobbes erroneously speaks of the selfish bestiality of human beings and their need for an externally imposed social contract. Our basic innate capacities are, Roy Baumeister would assert, for caring, helping, and cooperation, with selfish and deceptive variations, of course.

Keep in mind, however, that innate tendencies can be molded, stifled, or exaggerated by shifting social structures and power, evident too in chimpanzee behavior. According to de Waal in *Chimpanzee Politics*, research and observation demonstrate that chimpanzees value social relationships. Males in power positions are less likely to intervene in disputes because of personal preferences. Sympathy based coalitions and interventions occur more frequently and commonly among females and children, and less so with high-ranking males. There is a type of moral scope, if we can use that expression, among chimpanzees, where good behavior is rewarded and bad behavior is punished. Again, though, we should note how de Waal would be the first to emphasize our connection not only to the chimpanzee but to the more loving and amenable bonobo.

Krebs outlines the development of the moral sense according to Darwin. The terminology "moral sense" had been used by the eighteenth-century British moralists David Hume and Adam Smith and refers to physical sensations of approval/disapproval and feelings of good/bad and right/wrong that are negotiated in our brains. First there were the pro-social instincts of primates and ancestral human beings, followed by the conscience. Later developments, says Krebs, because of language, means any pro-social instincts were concretized into *mores*. Eventually, *mores* were "strengthened by habit" (41).

While this seems quaint when so succinctly stated, it is valid since fundamentally Darwin saw continuity between sociality and the avoidance of others' disapproval. The early British moralists, Shaftesbury and Hutcheson, who anticipate Hume and Smith saw the moral sense as an approval function. Krebs points out that Darwin, no philosopher himself, has an ethos which combines Kantian rationality (ought) and Humean sympathy. There is descent from Kant, to G.E. Moore, to Rawls that leans toward an abstract ought above nature. See, rather, how Gazzaniga, in *The Ethical Brain*, shows that we are neurobiologically hardwired to be ethical, although what we call ethics can of course vary from one culture to another. One's needs and desires, or a group's, are always offset by another's. How does one respond? Frans de Waal has done quite a bit of convincing primate research in an attempt to place the emphasis on the side of an instinctual empathy, which is rationally balanced by the prefrontal cortex in human beings.

The difficulty with any sweeping evolutionary story of morality is that it does not account for individual influences or the power of the individual to affect and alter the group. Psychologist Jerome Kagan concludes that despite social-environmental factors, individual temperament persists over a lifetime. Darwin focuses, too much, Krebs believes, on cooperative behaviors among many and fails to consider fairness and reciprocity, which can be highly individualistic and operate between just two persons. Darwin also places emphasis on human reason in morality, but it is more accurate to say that any moral sense, a sensation, relies on understanding and emotion. This is why, I argue, individual artistic culture intervenes powerfully in the social sphere.

Moral Origins and the Individual

Many scientists, from Paul Ekman and Joseph LeDoux to Antonio Damasio, have placed emphasis on the role of emotions in reasoning. Joshua Greene and Marc Hauser have written about the degree of cognition in moral behavior, Greene seeing more cognition involved in these processes than Hauser, though we are now splitting hairs. Darwin leans too much to the consequences or common good of moral acts and ignores intentions. Certainly this is a complex area, and even Adam Smith, before modern psychology, noted that not only could one have conflicting motives before acting, but one could also have bad intent and yet seem to act with care.

Primitive behaviors are covered by Krebs in terms of hierarchy, self-control, altruism, and cooperation. Without stating anything with absolute certainty, Krebs suggests that "perspective taking and moral reasoning" probably evolved in primate cultures (75). Certainly we see this is true in work on great apes by de Waal, Goodall, Fossey, and Galdikas who conclude that these higher primates rely on hierarchy and status as part of their social structure. Fighting for a resource is risky and a waste of energy, to say nothing of the fact that one might not gain the prize. Therefore, forms of cooperation evolved.

We see the origins of deferential strategies where resources are relinquished to the more powerful or one of high status. Krebs says that those in high ranking positions have greater degrees of neurohormonal chemicals like serotonin, vasopressin, testosterone and are better than subordinates in reading minds. He cites work by, e.g., Christopher Boehm, who ultimately sees the spread of egalitarian attitudes in early human species in this regard. Self-control, especially individual differences in the modulation of desire and aggression, comes into play. In evolution the accretion of results is what matters, and it seems clear that behaviors selected for include self-regulation and cooperation. Individuals successful in maintaining those self-governing and monitoring behaviors would have had consistently greater mating choices and success than aggressive, lone-wolf cheaters.

In terms of aggression, Jane Goodall looks at hunting in chimpanzees. This behavior is opportunistic, spontaneous, and quite violent when the prey is meat like the infant colobus monkey. There is evidence of cooperation, and while chimpanzees certainly do share food, meat is so highly coveted that they fight each other to get it. Even a low ranking individual will cling to a kill. Concerning other aggressive behaviors, Goodall says that temper tantrums are not uncommon, and aggressive behavior could be for punishment, coercion, retaliation, the release of "frustration," challenging, "intervention," protection, support, or attack on a conspecific (323–328). Male aggression is typically, but not always, related to the maintenance or achievement of dominance, whereas female chimpanzees will be aggressive for plant feeding and protection of young.

In terms of altruism, the driving forces are sexual selection, kin selection, and group selection. Darwin, for example, emphasized reproductive adaptations over those of survival. Since so much care is involved in raising young, early human beings found attractive those who would help both mate and offspring. E.O. Wilson in *Sociobiology* really framed the question by asking how we can account for altruism (giving) in light of natural selection (fitness). Why would I want to help you if there is no benefit to me? While there is, at base, an investment in genes, there is no analysis when it comes to altruistic kin behavior. Support of kin is spontaneous and emotional. Krebs notes that over time in spite of group against group, altruism will succeed over purely exploitive behaviors. Others might say that on any level altruism is really only a form of kin selection.

Altruistic behaviors like sharing, flexible tit-for-tat, or mutual aid ultimately benefit the individual and the group, so those less fit will observe and copy by deference those in higher ranks who most likely exhibit altruistic behaviors, even if the motive is for self-gain. There is, for instance according to Gilbert Roberts, competitive altruism for raising status. Those in higher ranks will tend to maximize their fitness, as natural selection demands, through cooperative adaptations. Altruism and social emotions could spread since high status individuals are quite visible and others will tend to mimic their actions in order to achieve, realistically or not, a similar level of fitness and status. There is, nonetheless, some evidence, Robert Wright reminds us, that pecking orders exist. Those in lower ranks are aware of their limits and will not attempt to rise, so that some version of tit-for-tat could occur between any types.

However, even though altruism and cooperation did and still do spread, there are those who will cheat because of selfish tendencies, lack of self-control, or low status. Bad behavior can spread in a group if all ascribe to some form of maliciousness. Yet a cheater only calls attention to himself and is usually punished, which then highlights the benefit of cooperative behavior. The upshot is that what we call morality is a social behavior based in emotions. Morality in this light is integral to art and cultural practices.

On sympathy, there are some who see positive social emotion as self-serving. Krebs says that some studies have pointed to sympathy as its own end. Perhaps. Even cooperating or sympathetic behavior that seems wholly unselfish might come with some unconscious expectation of a future gain in resources or a slight boost in status through the grapevine of gossip. Krebs acknowledges that social emotions, including forgiveness, evolved because of their inclusive fitness functionality. According to Hamilton, adaptations evolve because of inclusive fitness or the ability whereby an individual contributes to the reproductive success of a genetic relative.

In discussing human social behaviors, compared with other primates, Krebs notes that what makes us distinctly human is our ability to "show deference to abstract ideas...suppress selfish behaviors, control aggressive urges, plan for the future, and delay gratification" (163). These qualities are uniquely combined in us among primates but can vary widely in degree among individuals. The amount of individual difference matters since morality is dependent on social interactions. We have always been part of some group, but whereas in our ancestral environment the cluster was small, later tribal, now it has ballooned from towns and villages to cities and metropolitan areas. How much of the prehistoric forms of group behavior still linger in us to affect our actions? We routinely see how some small groups seek to overturn the status quo or inflict terror.

Regarding our prehistory, Alan Barnard says that hunting and gathering, but not farming, are adaptations to severe conditions, whether in deserts of Africa or Australia, the frozen Arctic, or the jungles of Southeast Asia. There is little hierarchy or class structure among hunter-gatherers and any stratification is based on sex and age. Any "sexual differentiation" comes in "subsistence activities" (55). Food sharing occurs so as not to aggrandize goods, since there is frequent movement. Any accumulation spurs an "obligation" to distribute materials according to certain rules (57). Hunter-gatherers do not have, as do agrarians and herders, a divisive us-them mentality but more of the cooperative you-I pairing and mind-sharing.

In *The Biology of Moral Systems* Richard Alexander says that indirect reciprocity or "cooperating with cooperators" evolved from direct reciprocity. Indirect reciprocity between different groups flowed from the biological processes that permitted exchange between group members. As indirect reciprocity spread, so did one's tendency to monitor very carefully any interactions. More to group selection thinking, Peter Richerson and Robert Boyd would say that there are two types of social instincts we evolved: the individual and the tribal. A form of cultural selection plays into the tribal mode. Without discounting other theories, Krebs seems to favor Alexander's, where the basic unit of selection is the individual, and so great emphasis is placed on the mental function of imagining and maneuvering, even playing with, a symbolic self among others. We can see how these early mental machinations over time and given the use of symbols became art cultures.

In a large group, the common interest served will suppress the selfish outrider, but who will do so and will that person be rewarded? Krebs notes that Christopher Boehm's account of hierarchy deals with dominance and is externally oriented. Alexander's account deals with small groups that benefit from direct reciprocity of mate and kin. Alexander's main thinking is that direct reciprocity became indirect, involving three parties, as small groups became larger tribes, and this behavior spread as more observed its benefits. The problem here is that even today indirect reciprocity is ripe for cheating. At the same time, one who acts fairly and honestly in third party exchanges on a regular basis advances his or her status and prestige across groups, and that good behavior gets noticed and copied.

Krebs sees reciprocity as individual selection and later cultural. This type of interaction among people is part of the so-called social intelligence theory where one began to picture and guess at the intent and motives of another. There is a cause/effect relation between the human brain and the ability, beyond social emotions, to make reasoned moral decisions and reflect in conscience. So it is important to stress something missed by the naysayers outlined above, namely that reason evolved to solve adaptive problems and not to puzzle out free-standing, metaphysical abstractions about Truth or Infinity in the universe.

Krebs is clear about how perspective comes into play in any discussion of the moral senses. Typically, we do not feel guilt for another's action and we do not experience moral outrage over our own bad actions. Pro-social behaviors gave rise to duty, ethics arose from norms, and conscience grew from "emotional reactions" witnessing the antisocial action and subsequent punishment of others (203). Rather than saying *moral sense*, Krebs says *moral senses*, since there are approval/disapproval mechanisms of *feeling* and of *thought*. These mechanisms are in part directed inwardly, in part outwardly, sometimes including thoughts and feelings prior to a decision, and sometimes only upon reflection.

To complicate this picture, Krebs notes how we can and do form moral judgments in other ways beyond personal/social emotions and group norms. Namely, we tend to moralize by another's rank, accrued deeds, and even overall hygienic appearance. Where is rationality in this type of visceral evaluation? We can see how there are visual and other sensual aspects related to what we term morality. Our moral judgments can be colored by our location and its relative cleanliness. This is not merely approval/disapproval but an evaluation as to the perceived intrinsic worth of something and how it would adaptively help one or not. Some of these moral evaluations stem from our long past and can be seen in our primate cousins' disgust of filth and distaste for strangers, as noted by, for example, Goodall and Fossey.

Krebs reports that cross-culturally, human beings categorize social behaviors as "affectionate," "hierarchical," "egalitarian," or "economic." While chimpanzees are capable of the first three, only human beings can combine all and especially utilize the fourth (220). Our abilities to cooperate mean we will alter and even change our beliefs to do so, and these types of mental adjustments in social interactions would account for the development of theory of mind. Modifications are also implicated in culture and artistic behaviors. Once again deferring to the individual, Krebs notes that while there are universal norms, these norms evolve from the types of moral evaluations individuals decide to make and keep.

The origin of social *mores* is not in the environment per se, but the environment of learning will help spread *mores*. So while there are innate, genetic dispositions that generate a moral culture, that very culture is an essential prop for the maintenance of norms, unless it is a purely universal norm, like fairness. Culture depends on how any group has decided to solve an adaptive problem, and hence why the universal notion of sharing will differ in particulars among cultures. We see, too, widely different applications of justice from one culture to another, from one country to another, and even from one state to another in America.

Krebs cites research that suggests Utilitarian models of morality are cognitively driven whereas deontological models of morality are emotionally driven. But this type of split is only superficial because we originated it. Human beings "are naturally disposed to help others... [and] to obey rules..." (245). Since these tendencies emerge in a social environment, Krebs is keen to recognize and not dismiss social learning, another of our adaptive functions. Biologists like Eva Jablonka and Marion Lamb include in social learning organic-related, environmental elements like epigenetics (external influencers on how cells work) and even transmittable behaviors and symbols.

Krebs returns to Kohlberg and, as reluctant as he is, criticizes data that was derived from studies based on hypothetical questions to only male students in a university setting. For example, he debates the use of why-type questions which do not reveal one's moral grounding but instead become an "intellectual exercise" (265). Krebs cites studies that demonstrate how the reality of a situation draws responses dramatically different from any theoretical question/answer. Even more difference occurs between men and women. Think, for a moment, about all of the ink spilled analyzing Philippa Foot's runaway tram scenario, and all of its variations, e.g., the footbridge. We will get to this moral philosophy exercise shortly.

Ultimately, Krebs says that while social learning is an important part of our cultural *mores* it does not account for the origins of moral systems. There are fibers of what we call moral behavior ingrained into our evolved nature. Staunch evolutionary psychologists Leda Cosmides and John Tooby would suggest that different modules are simply part of our evolved brain mechanisms, inherited from our ancestral, prehistoric past. These modules served adaptive, social challenges about, e.g., selfishness and aggression that increased fitness. Others, including de Waal, insist on a continuity of empathetic feelings between us and great apes. Whatever the case, there are visceral approval/disapproval emotions evident too in the making of and participation in artistic culture.

Krebs admits that putting together the puzzle about the origins of morality is difficult. Since his training was with Kohlberg, in spite of his insistence that he has moved away from him, we can still see his sensitivity to social learning although he places more emphasis on evolution. Some will say that evolutionary studies have nothing substantial to add to the origins of morality and that evolutionary ethics is theoretical. But the evidence is available and has been studied by too many biologists and psychologists to ignore. Biologist David Haig, for example, suggests, first, the "social gene" and then, later, "intrapersonal reciprocity," where, despite conflict, internal gene equilibrium is reflected in external social equilibrium.

Others might dismiss evolutionary studies altogether, believing that every other part of us could possibly have evolved except for our brain, and that our ancestral environment whose remnants are evident in other behaviors has nothing to do with our moral sense. This thinking is patently false. Like our morphology, the brain is an adapted mechanism for complex social organization and behavior. Brain structures and functions relevant to moral sentiments evolved to include many parts. For instance, as scientist Douglas Allchin notes, various brain areas enable us to see without judging moral images, to gauge another's intentions, to determine when any moral situation becomes personal, to abstract a problem, and to mediate conflicts. Patricia Churchland submits that there is continuity between the social and moral spheres, supporting her claim with studies in neuroscience showing how the same area of the prefrontal cortex becomes active whether one sees a social or moral image.

All of this is why I say there's a moral basis for artistic culture. My whole argument is that as social animals we have learned to use culture and art behavior to address moral issues. Art implies responsibility and the freedom to enforce or challenge cultural practices. Our autonomy is linked to our biological feelings, yet morality is an emotional behavior guided by reason.

Moral Origins and the Group

In his paper "Defining the Evolutionary Conscience," from which I draw in the next few pages, Christopher Boehm says that *conscience* is where moral behavior begins. While this paper precedes publication of Boehm's *Moral Origins*, it clearly advances ideas more fully developed in that book, which I will discuss. Boehm's argument leans toward group selection. Looking back Boehm characterizes our last ancestor shared with the chimpanzee as a hierarchical being who understood rules, dominance, and submission, therefore capable of forming coalitions. In making lesser coalitions early persons would attempt to minimize but not eliminate the powerful alpha male. This behavior becomes for Boehm "a critical preadaptation for moral origins" (28).

Prehistoric hunter-foragers would have shared meat, and divisions would have been made by dominant males based on kinship and alliance. Since chimpanzees and bonobos are capable of self-recognition, they could take perspective of others' attitudes, an integral ingredient to cooperation and deception. But Boehm is quick to note that these early people probably felt fear, not shame, the latter a key component to his theory. Nevertheless, public punishment in or by a group no doubt had real genetic effects, since today we see that both chimpanzees and bonobos can and will at times form a coalition against a high-status member of the group who is overly aggressive.

Boehm is convinced that social control of an overly dominant individual sets the stage for conscience. Boehm lays great weight on large game-hunting, which at 250kya with *Homo sapiens* was not as crude or haphazard as it had been previously. Human behavioral ecology, or how human beings optimized the environment, dictates that the reliance on large game evinces social organization and communal meat distribution to satisfy increased carnivory. A key caveat here is that in spite of any egalitarian food distribution, once the meat was in the possession of any family it was considered "private property," says Boehm (30).

Boehm is aware of critics who say there is no real parallel between prehistoric hunter-gatherers and those of today since the latter are few and far between and might have been contaminated by modernisms. Boehm's response is that Pleistocene foragers were already marginalized by severe climate changes which caused "serious subsistence stress" and probably enhanced "cultural flexibility" (32). Indeed, current hunter-gatherer-foragers are mobile and egalitarian, as prehistoric people would have been, citing Klein and McBrearty/Brooks. At a site in Israel circa 400kya, notes Boehm, there are carcass bones with "many individual cutting styles and many angles of approach..." indicating fierce competition for food, whereas by the Middle Paleolithic period there is, typically, evidence of butchering by a single hand (33).

In early forms of hunting there was no social shame, but over time as the strong alpha types, as Boehm characterizes them, suffered criminalization there was a fitness gain for those who controlled their aggressive urges. The mental and physical regulation of desires would have resulted in the inheritance of gene frequency changes toward cooperation. In this way, public punishment of one or a few by the group or by a coalition of groups would have set in motion what we now call conscience. For Boehm, conscience is both prosocial, as in self-control, and fitness enhancing, as in reputation. Boehm does not see any form of a social calculation as does Richard Alexander, though one wonders if Boehm is not being overly optimistic.

Because he relies on advanced group hunting which he sees as a later development, Boehm dates social selection at a late 250kya, followed by group norms and behaviors, and finally to individual fitness enhanced by social rewards. In this chronology, he talks of empathy, a word used by most social scientists and primatologists, whereas the active form, sympathy, might be preferred. At any rate, consistent altruism reward and free rider suppression on the individual and group levels began to fix and spread these genes in the pool of hominins.

Let's now turn to Boehm's book on this subject. Going against the grain of individual selection theories, which posit the emergence of altruism from parental bonds and kin relationships, in *Moral Origins* Boehm makes a powerful argument for group social selection to account for the advent of altruism. Paradoxically, according to Boehm, altruism occurred through negatives. There was punishment of free-riders and subsequently the fear of public ridicule or embarrassment. Boehm claims our moral origins lie in the adaptive design and flexibility to rules of the conscience away from "fear-based" bullying and its concern with maintaining personal reputation (176). Drawing from experience as a field researcher with primates, working with Jane Goodall, and from his research on Pleistocene-like contemporary foragers, Boehm concludes that small bands of people pressured others to act generously for the sake of group cohesion and cooperation.

However, even as Boehm admits, there is no single theory that will answer the conundrum about the origins of human morality. Aside from their different approaches to charting the birth of moral systems, both Krebs and Boehm give voice to an exclusively evolutionary reading. And from these books, one can work backwards through the literature on this subject that started in earnest with Darwin in *The Descent of Man*. In our prehistoric ancestral human species and from the DNA level, the selfish-gene model is attractive. From the perspective of more recent history and the emergence of *Homo sapiens* and culture, the group model is attractive. Morality is a behavior that is expressed from a combination of factors, some at odds with each other. There is instinctual response to fear, selfish needs, and desires for mates or resources; then there is prefrontal cortex activity in terms of alliances, coalitions, and group homeostasis.

While he draws from some of the biological leaders in this field, including Trivers and Alexander, Boehm places emphasis on a social and not a selfish or kin model. Therefore, he often invokes sociologist Émile Durkheim. There is very little discussion of the biologist Hamilton, some reference to the political scientist Axelrod, and counter arguments to biologist George Williams, the last of whom argues that altruism evolved between individuals and is not a group product. Books by Matt Ridley, Robert Wright, and Marc Hauser are criticized for neglecting evolutionary history, which is not wholly accurate, since their discussions focus on evolution. However, Boehm takes pains, in the tradition of paleoanthropologists like Richard Klein, to chart human prehistory. Unique to *Moral Origins* is how Boehm meticulously links our ancestral past to our present and accommodates both industrialized and still-present foraging societies.

Boehm is a cultural anthropologist, so it is not surprising that he would lay emphasis on the group. Culture can be problem solving, and morality can be a group concern. Consider how Krebs, a psychologist trained by but then turning away from Kohlberg, lays emphasis on the individual. There is indirect reciprocity, and so morality is a personal action. According to Boehm, Richard Alexander toyed with group selection theory. Interestingly, Alexander, a biologist, is used differently by the psychologist Krebs and the anthropologist Boehm. Nevertheless, as Alexander himself proposes in *The Biology of Moral Systems*, there cannot be any altruism without selfishness. Both the punishment of and aid to another are motivated in the short or long term by one's self-interest. Matt Ridley in *The Origins of Virtue* insists for many sequential pages of explanation that while altruism is evident on a group level, the ultimate cause of altruistic behavior is selfishness.

In a nutshell, here is Boehm's line of argument. Easily by 250kya large game hunting of horses and antelope was done in egalitarian groups. Cooperative hunting replaced the alpha male hierarchy, which set the pace for altruistic sharing and meat distribution. A crucial development was the punishment of cheaters who demanded or stole more than their share. Boehm admits that hunting goes back much further, that there was large game hunting as far back as 400kya. Regular control of fire dates to 400kya. By 250kya large game was hunted routinely and butchered systematically. This is not a new argument. Ridley as well as Klein date hunting and butchering back to approximately 1.4mya. Novel here is Boehm's insistence on a complete shift to group culture. However, in *The Origins of Virtue* Ridley cites anthropologists Hill and Kaplan who say, "Societies…do not have needs, individuals do; and societies are the sum of individuals, not entities in themselves. Therefore, only by understanding what made sense for the individuals would anthropology make progress" (99).

Debunking the tolerated theft theory offered by anthropologist Nicholas Blurton-Jones, Boehm asserts that cooperative hunting and sharing would promote "social bonding," encourage "sympathetic feelings," and involve some form of "perspective taking" (139–140). One could argue that, nonetheless, there is always risk calculation in sharing. The fascinating aspect of this book is how Boehm correlates his theory to the many contemporary late Pleistocene-like communities he has assiduously researched, from the Inuit to the Kalahari. Boehm notes that early civilizations, and even contemporary foraging communities, employ "preaching in favor of altruistic generosity" (191). Preaching might underscore, however, our innate selfish tendencies that repeatedly need adjustment. The preacher, too, might be an individual within the group who wants his ego to supplant many others.

And yet intriguing is how Boehm uses evidence from our prehistory to bolster his message. The conscience evolved through a process of social more than natural selection as hierarchical coalitions were formed and it was paramount to choose "useful partnerships" wisely while sometimes severely punishing others (149). Boehm says that this idea of deviant punishment affecting gene pools and leading to a conscience is evident in Darwin and Trivers, the latter of whom suggests there is moralistic aggression. If the cheater is quiet there is no expression of the selfish gene, so that particular behavior might not genetically pass on. In his favor and to his credit, Boehm is optimistic and not cynical, and while we have evolved to be, generally, a caring and helping species, there is still an extremely strong, innate selfish urge that requires emotional and mental regulation.

Any egoistic gene/behavior or tendency does not disappear. In fact, we are prone to selfish behavior and hence the altruistic preaching. Care for others redounds to the self, either via kin or group selection. Altruists are universally compensated in some way, and yet after many thousands of generations we still see cheaters, deceivers, free-riders, and other forms of selfish behavior quite often. This is to say nothing of whatever selfish urges lurk beneath consciousness. Evolutionary psychologists Mark Van Vugt and Paul Van Lange make the claim that we evolved cheater detection methods to benefit the group. But deception-finding is merely a mirror of one's selfishness and is like theory of mind. These mental calculations are advantageous to the individual. The logic is as follows: Because we are self-interested we therefore know to doubt, indeed to question, the trustworthiness of another, especially if there is an outward sign of dishonesty.

Boehm's group model, like Robert Boyd and Peter Richerson's, might represent a conformist tendency dating only to the emergence of big-game hunter bands and does not necessarily explain the deeper origins of moral emotions. Readers can see, then, why I spend so much time charting human and ape correspondences. Outside of the group an individual would seek to conserve what little he has or has to gain for himself and his immediate family. Even at that, we are assuming parental investment by a male. Nevertheless, female care of offspring is ancient and neuro-chemically based. But within a large group the individual could attempt to exploit resources and profit by the expectation of receiving a share of winnings. In *The Cultural Animal*, Baumeister tells us there are gender differences in "social orientation," e.g., women more to intimacy and men to groups (110). Both are aggressive, but men are competitive in larger groups whereas women are competitive within family and smaller groups.

To some extent we need aggressive genes to keep us alive and going. In theory, the aggressive gene was modulated and never disappeared. Key to Boehm's thesis is that over time group suppression of cheating has raised the conscience to a level where *moderately* potential cheaters are kept in control. But for the most part, is this not via his reliance on Durkheim a form of the standard social science model, which Krebs de-emphasizes considerably? Our instincts and innate cognitive mechanisms ask that we cooperate in order to survive. No one doubts the practicality of the group, but we cannot over-credit the group for self-interested work of the individual. We know that cooperators prevail; see, e.g., work by D.S. Wilson. But the underlying motive for an individual to cooperate can be to second guess the needs and desires of his fellow hunters for self-profit.

Individual incentive to invest in the group can yield a more secure return, but this investment does not obviate the individual's egoistic needs or desires. With a nod to individual differences, Boehm admits that the free-riding gene is in the human gene pool but often suppressed at the level of the individual. Yet he insists that this selection occurs on the social and not individual level. Boehm dismisses as less responsible any kin-type, reciprocal altruism, or mutualism selection theories. Public opinion, gossip, and reputation selection undoubtedly impact fitness. Since many of our genes are tied to and expressed in the brain, there is a strong neural as well as environmental factor involved in behavior. Consider how Constantine Sedikides writes about the symbolic self in our prehistory in nascent form perhaps as far back as 1.8mya. That symbolic self would account for perspective taking and the importance of a public persona before big game hunting.

Boehm spends a considerable amount of time discussing Alexander. Whereas Alexander emphasizes *good reputations* as part of mating/cooperation, Boehm stresses how bad behavior will ultimately lead to gossip and then group punishment. Through the process of natural selection, public reprimand would have led to a "'debilitation of aggressive responses'" and a "'strengthening of inhibitory controls,'" quoting himself from *Hierarchy in the Forest* (167–68). Boehm is describing the evolution of conscience, which is like a "social mirror" reflecting our behavioral accounts, good and bad, for us to view in full (172). Without addressing brain science or consciousness fully, this is where individual differences come into play. Undoubtedly there have been and still are people who use this *social mirror* in a calculating manner to subtly deceive while appearing good.

Even Adam Smith in the eighteenth century, with his notion of the *impartial spectator*, recognized individual differences in the competition between caring and personal gain. Though as a product of his time Smith chalks up these differences to class and not to neural genes. At any rate, Boehm admits that since the tendency to altruism is slight, the hunter-gatherer groups he examines prove that culture is necessary if the group is going to survive without serious conflicts. For instance, in discussing tit-for-tat, Boehm says that the exchange of goods is less important than the "spirit of generosity" the exchange produces (302). Granted, but one knows that if he boosts the generous spirit of the group he stands a better chance of gain, for without any likelihood of eventual profit a player is sure to defect.

We evolved away from the hierarchical model to the egalitarian. More precisely, Boehm is able to delineate how and why human conscience arose. Conscience is more than the function of the individual in a group and more to the function of the group on the individual. Any social hypothesis for the gradual increase of moral behavior seems to ignore the fact that our deep genetic history is one of self-interest. Outwardly we became more cooperative and less aggressive, evident in morphological changes like minimized canines and reduced male-female dimorphism. But internally our cognitive functions increased in terms of theory of mind, anticipating another's intentions, deception, and cheater detection.

In spite of facial expressions, any adapted mental functions are hidden from view, and there is no accident in our having evolved private thought processes.

Moral Feelings: Ape and Human Continuities

In *The Bonobo and the Atheist*, Frans de Waal confirms the pro-social behavior of chimpanzees and bonobos. These great apes express emotions, engage in communal activities and tasks, present and maintain codes of behavior, and demonstrate care and concern for others. Meantime these apes evince individual personalities. Mustering his own research and work from others, one of de Waal's repeated points is that primate altruism need not involve cost and need not be reciprocal. Rather, he emphasizes how altruistic behavior is based in empathy and is automatic, which would make it evolutionarily old and genetically heritable.

The thrust of de Waal's claim is that our moral behaviors are rooted in nature and come from within us through our adapted prehistory and continuities with primates. What we deem moral and ethical are not special designs handed down from an outside source. Instead, caring, empathy, approval and disapproval similar to our own are apparent in species close to us genetically and evolutionarily. While de Waal is critical of millstone-like religious codes, he maintains against atheism that we have evolved religious beliefs and values for a reason.

As in his other books, including *Good Natured: The Origin of Right and Wrong in Humans and Other Animals*, *Primates and Philosophers: How Morality Evolved*, and *The Age of Empathy: Nature's Lessons for a Kinder Society*, de Waal's theme is that we are overwhelmingly, though not always, caring and cooperative creatures. We have evolved pro-social tendencies as the evidence of similar traits in our living primate cousins supports. This focus on empathy is evident, too, in work he has done with Jessica Flack and Stephanie Preston. De Waal marshals evidence of spontaneous *good deeds*. For example, there is one chimpanzee who, over a long period of time, helped an old, unrelated one walk and get water. Similar instances of *animal empathy* occur in dogs, elephants, rodents, and birds.

My attention in the next few pages will focus on *The Bonobo and the Atheist*. In terms of the philosophical and religious undertones of the book, de Waal is skeptical of the so-called detachment of Adam Smith's impartial spectator. De Waal is critical of Christopher Hitchens' atheistic Marxism, Sam Harris' atheistic reason, and Richard Dawkins' atheism which assumes that any religion is a dangerous delusion. De Waal is a bit softer on Daniel Dennett and credits him with at least bringing up the evolutionary importance of religion. Nicholas Humphrey, in discussing the social function of the intellect, has called religion a way for early human beings in sacrifice and ritual to *bargain* with an often hostile nature.

Utilitarianism has no place in de Waal's one-on-one morality. Peter Singer is a contributor to *Primates and Philosophers*, and we will discuss Utilitarianism in a moment in looking at Greene. B.F. Skinner comes under question several times, but this is not surprising since Freud, Kohlberg, behaviorism, and the standard social science model have been eroded by evolutionary psychology. Skinner did not consider emotions connected to behavior and believed, rather, that all behavior was molded or modified by repeated reward or punishment conditioning. We know now that much behavior is innate. Consider, e.g., Paul Bloom and the innate moral life of babies, or Steven Pinker and evolved cognitive functions.

As Hume famously puts it, reason is a slave to the passions. And de Waal says that contrary to Hume, i.e., Kant, moral philosophy is misdirected in its top-down approach. He correctly reasons, "Imagine the cognitive burden if every decision we took had to be vetted against handed-down logic [reason]" (17). We have instinctive moral emotions that have guided us, before codes and rules, about which behaviors to approve and which behaviors to disapprove. In addition to persuasively arguing for moral sentiments that are products of natural selection and are continuous with primates and to some degree mammals, de Waal questions why atheists comprising Dawkins, Hitchens, and Harris *insult* religious people in the name of science when scientists typically are not interested in dictating human behavior.

Far more interesting, as per Barbara King and Pascal Boyer, according to de Waal, is to understand why we have religious beliefs, values, and rituals in the first place and why they have persisted. In terms of adaptation, our ancestors encountered confusing and contradictory data inputs that could be resolved or mitigated only by manufacturing supernatural beliefs. These behaviors evolved into ritual and religion with more complex societies, suggests Scott Atran. Who can disagree with de Waal's assessment about atheism: "the (non) existence of God…strikes me as monumentally uninteresting" (21).

De Waal, understandably, is incredulous with atheist *rampaging militancy* in light of the fact that they are literally fighting for nothing. He cites Stephen Gould and his notion of *magisterium*, where "science and religion occupy separate spheres of knowledge," the one dealing with the stuff of the world, the other dealing with our existence in the world (105). De Waal strikes out at the militant atheists, insisting that the "enemy of science is not religion" but is dogma over intelligence (109). Rather than ignoring beliefs and values canned as religion, de Waal wants us to consider adaptive functions.

According to de Waal, genetic information suggests that chimpanzees have changed over time more than human beings. We split from the chimpanzee via an unknown type, but surely an ape. For instance, there is *Ardipithecus ramidus*, one of the oldest known hominins according to Klein, which is less chimp-like. De Waal wonders why paleontologists have not traced our descent from a bonobo stock, which bears resemblance to *Ardipithecus ramidus* in its less developed canine teeth. De Waal says that the chimpanzee, therefore, might be in our diverse and divided descent as we can so far chart it, "a violent outlier in an otherwise peaceful lineage…" (61). This is not to say that bonobos cannot be violent, but the cases of killing documented in the wild and in captivity are much fewer compared with lethal incidents among chimpanzees. Bonobos have serious conflicts but make a concerted effort through sex to reach reconciliation, and while they can be aggressive hunters, it is for small game of another species.

Our human ancestry based on the most current DNA genome project puts equal weight on bonobos in our heritage and not full weight only on chimpanzees. Though there is a common ancestor, bonobos were cut off geographically by the Congo River from chimpanzees and gorillas, a significant elimination of aggressive competition. Bonobos share more recently, 2mya, ancestry with chimpanzees. But was that ancestor more chimpanzee- or bonobo-like and from which human beings derive? The answer to this question is more paramount than locating the so-called missing link. For instance, not only is the distress recognition area of the bonobo brain "enlarged," its brain also has more developed neural patterns that would be implicated in controlling aggression, according to de Waal (80–81).

Related to this lineage and to social behaviors Jane Goodall reminds us that chimpanzee society in the wild is fusion and fission, where individuals are quite autonomous, mobile, and free. They are often not together, either in ranging, foraging, or sleeping. This free movement is in contrast to baboons and mountain gorillas, so the same chimpanzee individuals might not encounter each other on any single day. Hence there is doubt in terms of who will meet whom, and all of this ambiguity in handling a fluid social structure indicates high intelligence. Most relationships among chimpanzees change over time as they develop physically and mentally, though there is relative stability with kin (e.g., two brothers). Therefore, Darwin's branching idiom is crucial when considering our deep history, since we share traits from a number of ancestors. As should be evident to my readers by now, there is no linear development of *only* a human species.

Citing research by psychologist Ulf Dimberg, not published until 1990 because of resistance to its findings, de Waal strongly suggests that empathetic responses are unconscious and do not require deliberation. This means that empathy is a response to "bodily connections involving faces, voices, and emotions" (132). Empathetic response is apparent, based on de Waal's research, in apes, and there is "no sharp dividing line between human and animal emotions," as experiments with mice and rats attest (137, 142–143).

This is not to assert that thought processes are completely absent from helping behavior. We have a first-response visceral (limbic) brain and an executive functioning neocortex. De Waal has four thousand observations of chimpanzees and says that often help is given through a combination of empathetic emotion and a cognitive filter.

At the same time, chimpanzee society is quite hierarchical socially in terms of who is the first to get food or to mate. De Waal characterizes hierarchy as inhibitory since one is required to control his or her needs and desires. De Waal suggests that archaically we had a similar system, which ultimately helped us evolve morality. This moral system is built not on instincts, which are fairly consistent and somewhat inflexible, but on variable emotions. Primates and certainly human beings are capable of, in many cases, deliberation. So de Waal sees that the social system is *reinforced* in two ways. First, we operate internally, with empathy, to avoid conflict. Then we respond externally, with punishment, to correct violations to the group norms.

These two underlying struts are connected and make sense, says de Waal, from an evolutionary perspective since each individual desires "integration" and not "isolation" (161). At least for human beings, very often a decision is made emotionally and only later justified by some rational explanation. Although we are motivated, for the most part, by needs and desires, and not exclusively by reason or abstract ideas, we exhibit on a daily basis an extraordinary degree of self-control. And so do apes. Unlike apes we can channel our emotions, needs, and desires in creative ways, and so we have culture and make art. De Waal notes that great apes "respect each other's possessions..." to such an extent that on many occasions he witnessed high-ranking males beg for food, rarely using force (161).

The upshot, since we know apes grieve and have rituals, is that there is evolutionary evidence for a biological origin of morality. By studying our ape cousins we can step back in time. Morality is a case of the basic biology of survival and reproduction observable in the natural world. These basics include across many species, de Waal notes, "The desire to belong, to get along, to love and be loved...to stay on good terms with those on whom we depend" (228). The complex social interactions, reactions, and relationships we see in the animal world are not based on metaphysical rules imposed from above but have evolved from within the animal kingdom across many species, and that includes *Homo sapiens*. There is a basic moral system evident in the animal world, i.e., fairness or what *we* call justice.

Martha Nussbaum offers an intelligent and timely writing on human and animal compassion though light on any Darwinian perspective. Nussbaum says that human beings very often fail at moral compassion because of what de Waal calls anthropodenial or the human tendency to deny its own animal nature. Paraphrasing from Galdikas, we learn that soon after Darwin Westerners developed almost a hatred of apes, most likely since human idealism and Christian anthropocentrism were threatened. Apes were portrayed as monsters and routinely slaughtered. Travel accounts only perpetuated any misconceptions, fueled later by images from the 1933 movie *King Kong*.

While de Waal writes about animal-human continuities, Nussbaum says that others frequently place emphasis on "good discontinuities," how we are morally superior to animals and thus cling to the old notion of the Great Chain of Being which places human beings near the top. Nussbaum notes there are three parts or judgments of compassion. First, there is recognition of the *seriousness* of the situation. Second, one needs to ascribe *no-fault* to the creature for whom one feels compassion. Third, one must imagine some identification for oneself of *similar possibilities*. These judgments are not standard operating procedure for a human being, who typically spends his or her day goal-directed from only a personal point-of-view. Nussbaum says that since we have invested our psyches in certain areas of concern, those are the ones that will rouse the strongest emotional response. But empathy is not enough for compassion, since one can enter into another's feelings to identify a weakness and do harm. By analogy, consider how a dramatic actor can manipulate his or her feelings.

Comprehending suffering is not sufficient, since animals and children are so capable. One must be able to ascribe fault, or not, and so understand the seriousness of the suffering and situation. Nussbaum cites three examples. 1. Mice in pain were perceived in distress by those who knew them, perhaps emotional contagion but no empathy or seriousness. 2. An elephant shot and left for dead by poachers is consoled by other elephants, perception of seriousness of the suffering animal. 3. A human man cries when he learns on television of a foreign boy's birth defect and eventual death. The man's dogs console him, but if he were crying over a gambling loss the dogs would still console him, so the man understands the concept of no-fault but the dogs do not.

Human beings have a great sense of their own frailty and mortality, and so too it seems some animals with a sense of self, e.g., elephants. Compassion from animals is more limited to instinct and confined by a narrow window of time and geography, whereas compassion with human beings, says Nussbaum, "is profoundly uneven and unreliable…" (136). Nussbaum asserts that the unreliability of human compassion stems from the human tendency to deny its animalism. We misapply fault and so ascribe animalism instead, seen in the bad attitudes of sexism and racism. One does not want to be, much less be perceived as, helpless. So in most societies there can be dominance hierarchies, in turn fueled by a sense of disgust about others, which Nussbaum says emanates from the reluctance to admit our own animalism and bodily functions. If we can minimize others to a state of brutishness, then they don't deserve our compassion.

Let's take this a step further and re-focus de Waal's lens with a skeptical stance.

Joan Silk examines the roots of pro-social preferences, going back to the oldest thus known hominin, *Ardipithecus ramidus*, one of the last common ancestors shared by apes and human beings. Silk reminds us that in the hominin line brain development has been more significant and dramatic than morphological changes, and through a process of neural accumulation we have, no doubt, the sudden cultural explosion in the last few hundred thousand years. Silk notes that comparative studies between apes and human beings are important to help us determine evolutionary changes only in the human line after it split from gorillas about 9mya and chimpanzees about 7mya. Following the idea of kin selection, altruism, for example, commonly appears in nonhuman primates, mostly in the form of grooming. Other types of altruism include, but are not limited to, warning calls, coalition creation, infant care by a non-biological parent, and the sharing of food.

However, Hamilton, Axelrod, and Trivers at various times have also discussed reciprocal altruism or contingent reciprocity where an individual can act cooperatively towards an unrelated individual if there had been help received in the past and even if help could have been inadvertent. To engage in contingent reciprocity, individuals need to be able to understand individual differences and to recall the past so as to behave accordingly. As with human beings there are limits to contingent reciprocity, as in one's inability to comprehend an altruistic interaction, prejudice, forgetfulness, or impatience.

An organism will always seek to maximize its own fitness advantage. If there is a needed benefit in high demand, competition ensues, which can result in a behavioral change. If the change in behavior is beneficial, it will produce an advantage that is inherited and so becomes biological. In a competitive environment free riders are easily detected and often punished, and studies reveal, notes Silk, that "primates…balance exchanges over extended periods…" (12). This exchange equilibrium is contingent reciprocity.

Certainly, for human beings to engage in social interaction requires mental abilities to understand another's intentions and feelings. Studies show, Silk goes on, that great apes and monkeys, similarly, are able to recognize others as well as kin and have a sense of personal history, thus knowing who cooperates and who cheats. Moreover, Silk says that chimpanzees utilize "visual gestures" and especially rely on "orientation of the face…" (13). There is also evidence that chimpanzees understand intentions since they strategize competitively. Nevertheless, these ape skills translate to the capacities in a human infant little more than a few years old.

In spite of much evidence from Frans de Waal about chimpanzee and bonobo empathy, Silk maintains that these results are anecdotal and often subjectively interpreted. For instance, in one of the most celebrated instances, the gorilla Binti Jua rescued a child who had fallen into her area. But Silk notes that this ape had been reared by hundreds of human helpers. Moreover, the ape had operant training in mothering, which included fetching a rag doll and returning it to her enclosure for human helpers, the very action she imitated in rescuing the child. With Peter Kappeler and others, Silk disagrees with de Waal's assessment that there are *quantitative* but not *qualitative* difference between human beings and apes.

Would Martha Nussbaum see any real or perceived behavioral gap?

Silk et al. essentially argue for human emotional uniqueness. Contrary to Silk, Andrew Whiten in "Ape Behavior" does not see any breach between human culture and equivalents among apes. Chimpanzees show variation, from tool use to courtship displays, says Whiten, suggesting our own variety has roots in the animal world. Studies show, says Whiten, that chimpanzees employ, and have a great capacity for, the transfer of culture socially. William McGrew, too, asserts that culture is not uniquely human. Surely there is species differentiation, as Darwin showed, but mixed up in arguments for behaviors uniquely human are anthropomorphism and anthropodenial.

In further challenging de Waal, who clearly assigns empathy to great apes and sees direct continuities between them and human beings, Silk says that chimpanzee food sharing, e.g., mothers to young or males to others, is not as straightforward as it seems and might not necessarily instance altruism. For example, Silk mentions the so-called tolerated theft theory where one might relinquish extra spoils of meat. Silk says that food sharing among chimpanzees, in spite of pressure from conspecifics, is less frequent if there is any type of physical implement that could mitigate giving. Of course food sharing is the result of underlying tendencies, like cooperation and tolerance, so the actual sharing itself might not be the best measure.

Research by Silk at two different sites on chimpanzees who knew each other shows that in most cases in spite of any barrier chimpanzees worked so that they would reward themselves as well as others with food. Silk says that pro-social behavior comes at little cost, and virtually no distribution preference, indicating that "other-regarding sentiments did not conflict with selfish motives..." (16). De Waal and Flack's research, in contrast to tolerated theft, sees any aggression not against the possessor of food but against the beggar. There is, similarly, as Jane Goodall would say, a keen regard for any chimpanzee in possession of meat.

Citing studies by Warneken and Tomasello, Silk says that while chimpanzees might be pro-socially disposed, food can alter this equation, since food is not always readily available though sharing seems a norm. Concerning the last common human-chimpanzee ancestor, Silk assumes that for competition among resources those beings understood the relationships between individuals, especially in terms of intentions and desires. Further contradicting de Waal, pro-social skills and any sharing tendencies do not, insists Silk, necessarily imply empathy. Relatedly, Silk raises the issue of altruistic punishment, where at a cost to himself one will punish another for the benefit of the group, but this trait is not often seen in nonhuman primates. While Silk acknowledges pro-social behavior in hominids related to human beings, she seems to fall short of endorsing and is mildly critical of the continuities presented by primatologists like de Waal.

In chapter thirteen of *The Chimpanzees of Gombe*, Jane Goodall weighs in on friendly behaviors of chimpanzees and has much to offer. Because of years of child dependency on the mother and attentions received, chimpanzees expect and require touching, embracing, and kissing, e.g. for reassurance under stress. There are psychological benefits, says Goodall, related to social grooming which evolved from the mother calming the infant. Concerning altruism, Goodall discusses two types of costs: immediate, or bodily harm, and ultimate, or degeneration of gene pool potential. Since male chimpanzees do not know their offspring, any help provided is altruistic.

Not alone, Goodall cites instances of chimpanzees risking and at times losing their lives to help an unrelated conspecific. Since altruistic helping is genetic regarding kin, there is no need for any other non-kin altruistic mechanism to explain it. That is, it's there already, and memory comes into play since chimpanzees remember who has or has not helped. While it is not altruism per se, chimpanzees demonstrate other helping and sympathetic behaviors, says Goodall. There are close relationships among friends, greetings, food sharing, distress aid, and adoptions. Care for the sick and dying occurs for kin in the wild but also for non-kin in captivity since they are raised together. De Waal and Köhler also acknowledge these helping behaviors.

Morality, Feeling, and Cognition

In "The Moral Instinct," Pinker believes, contrary to Batson and Neiman, that there are innate moral sentiments. Pinker opens with a question as to who is more moral: Mother Teresa, Bill Gates, or Norman Borlaug. Most people say Mother Teresa, though Borlaug saved over one billion lives through agricultural advancements to reduce hunger. Gates donates millions of dollars to fight widespread diseases in developing countries. On the other hand, says Pinker, Mother Teresa ran well-funded clinics but preached asceticism and practiced tough love more than supplying tangible help. Mother Teresa not only received wide publicity but has a saintly appearance, meantime Gates appears *nerdy* and no one had seen Borlaug since he worked in a lab.

Pinker says this scenario is a clear example of how we can be duped by, perhaps even deceive others with, "moral illusions" (60). Any notion of moral illusion is not wholly negative and not cynical since it points to how we prize what is good or right, especially in others. Pinker sees morality as part of universal human psychology. We know that murder is wrong and we know it is wrong not to punish serious moral infractions. Nevertheless, there are shadowy edges to punitive moral thinking and, as Batson noted, it can be turned on/off.

Consider the subtle but important distinction, following an illustration from Pinker, between a health and a moral vegetarian. Each one adheres to different moral rules to achieve the same outcome. Culturally, over time, moral attitudes can change, and Pinker notes those dealing with smoking, now moralized, and divorce and homosexuality, now amoralized. Following Haidt, Pinker defines morality along emotional lines, not rational. We react to a situation or action, and then we rationalize our response. In line with my argument, this moral rhythm occurs in the creation of and response to artistic culture.

Apparently, more parts of the brain are in play and almost in conflict in trying to resolve a dilemma that involves hands-on behavior than in simply calculating cost/benefit. I will cover this in the discussion of Greene. Pinker cites Haidt's cross-cultural work that demonstrates five evolutionary moral senses: harm, fairness, community, authority, and purity, though these too are subject to variation.

So, unlike Batson, Pinker thinks there is evidence for the existence of moral genes. For instance, conscientiousness and agreeableness, two of five commonly known character traits, are related in identical twins when separated at birth, and psychopaths possess a type of moral blindness from childhood. The genesis of this evolutionary tilt toward morality probably had its roots, according to Robert Trivers, in tit-for-tat trading without being cheated. But not being cheated also meant gaining the larger advantage, which led to competition for the partner willing to sacrifice the most. Competitive behavior led to offering the appearance of being the most generous and the fitness value of having a generous reputation. Ultimately these behaviors led to some people becoming, says Pinker, "generous and fair" (72).

Over time most organisms have learned that it is advantageous to act nonselfishly. But emotional self-interest can never be ruled out, as evolutionary psychologist Gilbert Roberts demonstrates in writing about competitive altruism.

Moral Dilemma and the Runaway Train

In *Moral Tribes: Emotion, Reason, and the Gap Between Us and Them*, Joshua Greene explains difficult and abstract philosophical ideas concerning morality from a cognitive perspective. Greene grounds his entire thesis around establishing a "metamorality" housed in Utilitarianism, which rubs against much of our discussion and its emphasis on individual thought and agency. But Greene is an important philosopher whose cognitive research is critical to this discussion. Greene points out how we often take a tribal stance by positioning our "interests and values" against those from another group (14). Throughout the book Greene strives to establish through Utilitarianism an over-arching morality that will attend to and care for all. But if our core biology and history are tribal, and if tribes have very different values, how is that going to change? The simplest solution to many of the world's problems is not to philosophize and seek a meta-morality that will cover all the tribes like a warm blanket but, rather, find a tribe that can act as a go-between for the dangerously competing tribes, a meta-tribe. We already have a super tribe, sometimes effective, sometimes not, in the United Nations.

Although natural selection equates to self-interest, morality (above moral sensations) is tied to cooperation, since through cooperation even selfish individuals can gain an advantage. The key point is that cooperation evolved as an advantageous contrivance among certain people in a group. Technically speaking a group can be small, but a tribe can consist of related groups. Greene's thrust is that our brains "did not evolve for cooperation *between groups*…" (23). Of course some groups who learned to compete advantageously with other groups had the upper hand, so that competition underlies cooperation. Those who cooperate can be more successful, and Greene suggests, in spite of his emphasis on tribes, that he does not endorse group selection. What we call morality might be in-group cooperation, which evolved to help one group overcome another.

Extensive research and experimentation show that we tend to be concerned about others, even in some cases strangers, to the effect that we exhibit sympathetic visceral responses to their misfortunes or misery. While we can and at times do help others, we'll do so if the cost to us is not great. These findings are groundwork for later parts of the book where Greene will attempt to convince us that the Utilitarian perspective is the best approach, since by helping others we create an overall better environment.

Greene recapitulates research from Paul Bloom on the moral life of babies to show how infants are capable of evaluating behavior by favoring cooperation and ignoring non-cooperators. Greene notes we use accents and other speech cues to make judgments about our willingness to engage the trust of others. We sacrifice to help one group and yet harm another group. However, simply because we can be tribal does not ultimately mean, Greene stresses, that we are "hardwired for tribalism" (55). There are a number of factors on the personal, parental, peer, group, and social levels that can influence the neuroplasticity of tendencies to adhere to a group. And indeed, in our prehistory we did not start out in tribes but in much smaller units.

Only the human brain has evolved what we label morality as a means to permit group-to-group cooperation. We have, on the one hand, emotions that motivate us to care for those close to us and yet, on the other hand, emotions that dispose us to avoid and even punish others, especially those we feel as uncooperative. Nevertheless, Greene says, we've also adapted feelings that permit us, for strategic reasons of cooperation, to forgive transgressors. Even as infants we tend to judge according to loyalty and reputation, how someone behaves nicely or not to another. We always know our reputations are at stake. Self-consciousness and embarrassment tie in with how we fit into a group or not. We are concerned about our own status and are free to punish those who do not reciprocate. So there are many traits that have evolved on an emotional and not necessarily on a cognitive level to help otherwise selfish creatures cooperate.

Tribalism or cooperative partnership is a given, evident from research and studies on human infants and monkeys. Hostility towards outsiders is part of the evolution of cooperation. At this point Greene seems to suggest, beyond tribalism and groups, members of a group will adopt the group culture, whether it involves charity or killing. He also seems to suggest that severe local issues like poverty, violence, and climate change could escalate into inter-country problems. The reason we have conflicts between tribes, Greene says, is that some societies favor individual rights over the group, some have an obscure honor code, and some value religious beliefs more than others. In spite, or perhaps because, of our inherently self-interested tendencies, our moral problems tend to escalate into an attitude of us versus them.

Clearly these last few paragraphs fall in line with the ideas of evolution and human culture argued so far. Artistic forms of culture are moral since they spring from and affect cognitive emotions of approval/disapproval. Let's see how cognitive emotions work.

Greene gets into a very detailed and rich chapter on the variations of the runaway trolley scenario. For readers unfamiliar with this moral problem, one can search for trolley at the Stanford Encyclopedia page http://plato.stanford.edu/. Essentially, the trolley problem involves the question of switching the track of an oncoming train to kill one person rather than five people, or pushing a man onto the tracks to stop the train and so save five other people. At bottom says Greene: "our intuitions tell us that the action…is wrong" (117). Yet we'd do it anyway.

Here is the crux of the matter. Emotionally, in what Greene calls our automatic mode, we know that we should not harm someone else. But on another level that involves higher cortical regions, in what he calls our manual mode, we understand that harming one for the greater good is not only necessary but morally justifiable. Furthermore, there is a difference between hitting a switch to kill one person to save five people as opposed to pushing one person off a footbridge to stop a train to save five others. Our moral intuitions make us reluctant to engage in physical force on a personal level to help others, but we will. With Utilitarian decisions there almost seems to be ventromedial prefrontal damage in that there is only cognition without feeling. The heart of his book is on what Greene calls our dual-process brain, for if we acted on instincts alone we'd not be able to think through alternative situations or scenarios.

By analogy, my argument all along has been similar. We have, often at odds, emotions and thoughts concerning approval/disapproval that have played into the creation of and participation in cultural arts and practices. Artistic culture is a means to freeze, to abstract, and yet to call action. Art creation and participation can be simultaneously innate and yet manually regulated. More on art in Chapter 4.

Greene does not investigate any chronology or archaeology in terms of how the dual process might have evolved. In our prehistory we roamed as hunters and foragers in small groups, and so the automatic mode of no harm. Only later, with large game hunting, were there tribes. Imagine a wild animal racing at a group of five hunters and you have the opportunity to stop the attack by diverting its attention to someone else, or even yourself. How did that so-called manual or Utilitarian mode evolve? As outlined in other parts of this book, the answer might have something to do with our impulse for social strategizing, a combination of putting our self-interests and how we appear in line with the group.

The problem is that a terrorist group can label as Utilitarian acts or arts (or artistic destruction) they perform in the manual mode. Tribal cultures are moral only because they demand approval or disapproval.

In the control or manual mode we are able to look in all directions, consider, and weigh options, as opposed to the automatic or instinctive mode which simply reacts quickly. This is not to privilege the manual mode over the automatic. We have evolved instinctual responses and have retained them since they serve important survival and social functions. Moreover, different parts of the brain balance decisions, so that while one might strive to act for the greater good, visceral emotions could interfere.

In terms of brain evolution and function as related to moral decisions, we can handily manage cooperation in a group but not between groups. Certainly because of kin altruism as per Hamilton and Trivers this observation is not startling. There is a conflict between what our viscera makes us feel and what our rational mind makes us see. Gastroenterologist Emeran Mayer, for example, discusses the biology behind our so-called gut decisions. For the Utilitarian, the collectivist v. individualistic thinking is supposed to result in doing what is best for all concerned. While Greene is very accomplished at explaining Utilitarianism with long asides on Bentham, Mill, and happiness, he should not assume that his readers will become Utilitarian. His entire discussion is filled with generalities about the group to the exclusion of the individual. Individual selection is crucial to Darwin's evolution.

There is a distinction to be made between the individual acting for the greater happiness of many and the more ancient notion of excellence (*arête*) where the individual strives to sharpen her own wits, intelligence, strength, or moral virtue and so be happy. This concept of excellence is essentially driven by individual character, not the group. Not everyone is equal or wants exactly the same things or types and quantities of, to use Greene's Utilitarian word, happiness. Additionally, from a non-teleological Darwinian perspective, there is no progress or goal to happiness, since everything in life is a diurnal combination of variation, competition, and inheritance for individual fitness.

Greene optimistically wants "to encourage people to behave in ways that maximize happiness" (163). This thinking is a recipe for disaster considering our inherent self-interest and competitiveness. Whose happiness? Greene says the Utilitarian ideal is *impartiality* and "avoiding bad consequences" (168). This idea reminds us of Adam Smith who paradoxically pits sympathetic caring (*Moral Sentiments*) against self-aggrandizement (*Wealth of Nations*). Greene goes on to say that happiness is the bottom line and should apply across the board. But it appears Greene's equation for happiness is in high moralistic terms – saving some lives at the expense of one – and not in the more routine, basic functions or character issues.

Greene says objections to Utilitarianism come from automatic settings. Surely, since we are first and foremost emotional beings. We do not act with reason but first react, and then later, quickly or slowly, employ other brain areas related to the abstract faculty we call reason. Schopenhauer, a strong reader of Hume, cuts across Western philosophy and says human beings are not rational but irrational creatures. We have moral emotions and moral senses which, depending on whom one reads, are either mental faculties or not. Our default mode is self-concern, excepting kin, but we can be sympathetic.

Yet even in deliberation we might think away or rationalize the needs of others, not favor them or their so-called desire for happiness. While Utilitarianism might make sense on some high, idealized plane, it is not working in reality, in spite of what Pinker calls, borrowing from the biblically-inclined Abraham Lincoln, our better angels. There might be less overt statistical violence, but that does not preclude aggressive thoughts or violent urges, to say nothing of an entire entertainment industry that thrives off our visceral desire to consume violence virtually through various media.

The Utilitarian says, "no one is objectively special" (204). If Saints Francis or Clare of Assisi were to be on the side-track, we should hit the switch to kill either one of them to save the gang of thugs congregating down below and getting ready to attack Albert Einstein? Utilitarianism seems too clinical, without gut. There is almost something superficial about Utilitarianism in how it minimizes a basic truth about being human. Not only does each one of us think he or she is special, but we most likely also hold some others as special. If according to Utilitarianism everyone is equal across the board, then why not kill the five and save the one? Where is it written that quantity trumps quality? This is why the trolley scenario is dubbed a moral dilemma and why Greene makes the distinction between automatic and manual modes of mental activity. We've referred to morality as a behavior, and so it is both an emotional and cognitive behavior.

Greene's findings outlined in his chapter nine are compelling in terms of killing the one person (the pushing scenario) to save five people. Thirty one percent approved. And yet if we have pretty much the same scenario but have the man fall through a trap door activated by a switch, reminiscent of the original trolley scenario, sixty-three percent approve. The killing of one person is collateral damage to saving the five people since the one man, even when physically pushed, is in the way (so to speak) for someone to save five others.

Eighty-one percent approve killing one person to save five people if they see the death of one man not as a means with the use of personal force but as a side effect, says Greene. We are "emotionally blind" but not "cognitively blind" since from our deep past our ancestors premeditated actions, including violence (225). At the same time, reciprocity is also in our distant past, dictating an emotional reluctance to engage in physical harm since any hurt might return to us. We are nonetheless blind, says Greene, to anticipated side effects of violence, an action that does not fully account for consequences, and can push a man off a footbridge in order to save five others. Is Utilitarianism moral blindness, then?

On a related note, in chapter seventeen of her book, Goodall, tells the horrifying story of how one chimpanzee community systematically and with deliberation brutally exterminated another. The killing of conspecifics is uncommon among mammals since it involves great risk to the attacker. Since chimpanzees are aware of the feelings of others, and how another can experience pain, then with these actions they are aware of the cruelty they inflict, almost as if the fellows of their species are prey from another species. It is unclear from where our human tendency for violent aggression comes, and it need not be the chimpanzee line since bonobos are far less violent. Most likely, we evolved our own common tendencies for group-on-group violence. The point is that we can seemingly imagine away the life of anyone.

We sense harm to another as a means (pushing) but not so much as a side effect. We tend to be Utilitarian in more cognitively complex cases, where side effects are not necessarily visualized clearly, as in flipping a switch, opposed to simpler and more straightforward causality, as in pushing a man onto the tracks. Greene refers to the emotional response in these cases as an alarm call, but as a Utilitarian he fails to consider the wide differences in amygdala reactivity, temperament, or sensitivity. In the psych lab these results are solid, but even Greene admits that a lab is not reality. Brain scan machines in a controlled environment indicate which regions get hot, but this does not measure, importantly, outcomes. Greene says that although we might push the man off the footbridge to save five people we *feel* that this action is wrong. We sympathize with and act on helping tendencies for people we can see or know, an identifiable victim.

As de Waal has demonstrated in *Chimpanzee Politics*, whatever ultimately happens on the outside in a group, we can be sure that there has been lots of internal and self-interested strategizing going on mentally in terms of key alliances, coalition formation, and striving for rank. Greene's says the more we think about a moral problem the more we tend to gravitate toward our core, tribal, biased beliefs. We tend to rationalize our behavior, and self-interested psychological posturing is not precisely moral. Tribal differences can increase since so-called *rights* are established and asserted at all costs.

Forms of these rights are evident in cultural practices and arts, whether generated by the individual or to uphold the status quo.

Evolutionary Philosophy and Human Morality

David Lahti tries to bridge the separation between philosophers and scientists. In "On the Partnership Between Natural and Moral Philosophy," Lahti attempts a rapprochement between the apparently distinct disciplines of philosophy and biology. E.O. Wilson has envisioned philosophy and biology not on separate planes, as some philosophers have asserted. Lahti echoes a point once made by Jerome Kagan. If asked to define what is moral, a scientist, social scientist, and humanist, especially a philosopher, will each provide a different answer. But Lahti does not end there.

Lahti sees a "fault line" in academia in this regard (229). Surely we have seen that some of the philosophers surveyed at the beginning of this chapter seem to take pleasure in jabbing at scientists more than how Socrates acted as the truth-seeking gadfly. Lahti's goal, and the view I obviously endorse, is for simultaneous *endorsement* of the biological and philosophical points-of-view. Some philosophers agonize over genetic determinism while biologists assume the ability to move moral philosophy into the area of applied science. Lahti says that in spite of its thousands of years of history, philosophy needs to make advances, as has science, in the understanding of our place in a natural world. There are *temporal processes* at work which philosophy fails to consider.

Here's a pertinent quote from Lahti: "Because of the paucity of our scientific understanding of ourselves until recently, philosophy has been granted nearly free rein over human nature" (231). Even that statement requires elaboration. From a Darwinian perspective the modern synthesis of natural selection and genetics did not begin to flower until the 1920s and 1930s, whereas moral philosophy, with reference to David Cooper's anthology *Ethics: The Classic Readings*, goes back, in writing anyway, to about 500 B.C.

From a biological perspective, says Lahti, a scientist asks why heretofore explanations of human life ignored or outright excluded most of the natural world. Why do such explanations seem as self-entitled abstract units holding their own with no empirical grounding or test? On the other hand, a moral philosopher would counter by saying that ultimate questions are at stake, and if what we call morality is not thoroughly examined from a long philosophical tradition then there is only a biased opinion. The multiple viewpoints on moral issues that science and social science offer are exactly what irritate some philosophers. As per G.E. Moore and Putnam, *moral goodness* cannot be identified as empirical or physical. Science ignores distinctions and assumes there is moral weight in a sentiment when there might be none.

Let's return to Sharon Street. She blurs innate, fixed patterns of behavior with morality and yet ignores comparative primatology. According to Street, the dilemma for Darwinists is to explain the relation between the forces of evolution and what Knut Olav Skarsaune says are "evaluative facts." These facts are values related to good or bad, reasoned norms, and ought. David Copp says the dilemma for Darwinists is to render as *plausible* evolved beliefs that are responses to these moral facts. Where do we find in Street's argument any evolutionary timeline? Where is the fossil record or artifacts? As Lahti suggests, we see the philosopher often looking for what Street calls an "evaluative truth" – an abstraction. I don't construct an abstract philosophical argument but usher a cohort of thinkers across chapters who reasonably demonstrate how Darwinian moral realism is plausible, and I ground that plausibility in primate emotions of approval/disapproval later connected to material culture.

Morality is individually modulated emotional and cognitive behavior that links us to primates, especially to great apes. Caring, altruism, reciprocity, and moral behaviors are grounded in individual visceral responses more than in reason. That statement, however, is not meant to discount the significance of our rational mind. Since moral sensations deal with approval or disapproval, we can also find moral behavior associated with groups. Cultural practices, for example, exhibit many moral emotions in either upholding or challenging the status of a group. As I hope is clear by now after Chapters 1 and 2, many attitudes, beliefs, and values, and by extension artistic culture, are an amplification of evolved and inborn moral sentiments.

Let's now look more closely at the evolution of culture itself.

CHAPTER 3

Culture and Evolution

Cultures and sub-cultures account for the wide variety of differing values and beliefs. Roots are in part evolutionary, but social learning can heighten, diminish, or reshape natural tendencies. Pioneering primatologists including Goodall, Fossey, and Galdikas demonstrate that great apes have a form of culture, broadly speaking. Like us they act on instincts, but their individual and group behavior can vary widely from those inner drives and change over time. Not surprisingly, Goodall, Fossey, and Galdikas are not simply primatologists but conservationists, critical to this discussion. Just as we have seen throughout recorded history one human group aggressively eradicate another's culture, so too with human beings over great apes. We encroach upon their territories and eliminate them and their way of life. Many suspect this is how we supplanted Neanderthals.

As I argue in this book, culture is a moral issue. Culture, in line with what William McGrew suggests, is an adaptive, learning phenomenon. There are dimensions of approval, disapproval, good, bad, right, and wrong. Perspectives differ from one group and one culture to another, sometimes with baneful results. Small groups even today seek to eradicate another group's historical and cultural icons. We have widespread forms of culture in a high degree because we are socially complex. Intricate sociality is compounded by written and unwritten *mores* and explains why artistic culture is moral.

Roots of Culture and Art Behavior

Culture is of primary importance for human mental and social development. Howard Gardner, in *Frames of Mind*, indicates that were we not to have a cultural "symbolic code" our feelings, emotions, thoughts, and ideas would be mere ether (256). What I refer to in this book as artistic culture is the platform of our emotions and ideas. Selection can operate on cognition and behavior, and culture is evolved behaviors. The mind is not passive but adaptive so as to observe and conform or reject. Culture limits and binds. Everyone participates in and benefits from cultural productions. Culture means not just individual-to-individual but group and networked groups.

Culture broadcasts conduct. Where behavior is biological culture is a biological representation of individual or shared feelings and ideas. Culture is

 | DOI 10.1163/9789004319486_005

information, albeit recycled and re-modeled according to any individual or group. Culture can offer an entry point into something we've not yet experienced. Culture, however, permits an individual or a group to adapt, and hence why culture, biological at bottom, is also epigenetic (or how the cultural environment can act on cells). Rites of passage and rituals, for example, help individuals participate with, communicate to, and survive among others.

Eva Jablonka and Marion Lamb tell us that epigenetics is the inheritance or transmission of non-DNA information via cells. Cultural variation between groups is not necessarily genetic, initially. There could be evolutionary change over time as, in a process of natural selection, certain behavioral information is retained or eliminated. Natural selection presupposes that organisms are not passive but that their behaviors can have heritable outcomes. So, too, with artistic culture that is visible, heard, or tactile. Jablonka and Lamb say that like genes symbolic culture, as long as it is processed, understood, and retained, can remain dormant for generations before expression.

We are biologically programmed to seek and fulfill our own self-interests in terms of survival and reproduction. But we have evolved over the course of our environment of evolutionary adaptedness among other hominins complex social relationships that conceal, buffer, and mitigate self-interests. Jane Goodall reports that many subtleties of social awareness as well as appropriate behavior regarding aggression, alliances, and ranks, are learned from mother, group, and community. These latter two are fluid, and so in terms of intelligence they require close attention and excellent memory. We share these behavioral characteristics and traits, not surprisingly, with great apes.

Social awareness behaviors, inherited and in part acquired, translate to cultural practices like sociability, deception, manipulation, and hierarchy climbing. The forms of parental, familial, and social structures that contribute to human development are themselves cultural evolutions. These physical externals that push cognition are not free-floating reinforcers but are equally products of selection. In this way, genes and culture can co-evolve.

Close study of chimpanzees by Köhler, Goodall, and de Waal, and mountain gorillas by Fossey proves that these apes have distinct personalities. From cares and concerns, emotional expression and temperament, to needs and desires within a group, individuality is manifested in a social context. Galdikas, too, would assert orangutan individuality. More so because of their solitary nature, orangutans relate only on an individual basis and not to a group. Archaeologists Fiona Coward and Clive Gamble say that the way in which great apes interact culturally, in social situations and with objects, is similar enough to us, pointing therefore to our deeply shared cognitive history.

Chimpanzees are highly capable of noting distinctions not only among conspecifics but even in human beings. There is an incredible degree of attention paid to what others are doing and possibly thinking. De Waal notes in *Chimpanzee Politics* how apes are somewhat emotionally immature so that an excited chimpanzee in an aggressive encounter requires assuring and supportive contact. There is also much reconciliation between opponents in order to maintain key relationships. These are human traits too.

Although chimpanzees exhibit bias toward those they prefer to befriend, changes in coalitions among adult males determine power struggles and dominance. With further reference to *Chimpanzee Politics*, de Waal suggests that because of their immense ability at pretense together with their memories of prior experiences, chimpanzees possess what we'd call cognitive thinking. This mentality is done with purpose to create spontaneous, novel solutions to problems or dilemmas. The largest and the strongest chimpanzee need not be the alpha male. Rather, through a concerted series of monitoring, displays, screams, stomps, pilo erection, and psychologically calculated social manipulation, he can convey that image. There is a combination of his controlled deception so that the others make an *assumption* he's in power. This, too, is a human trait. In fact, one of the French editions of *Chimpanzee Politics* featured prominent politicians on the cover.

From these roots sprang some aspects of our own culture. As I have been arguing, in one way or another cultural behavior arises from our self-interest. An individual need not be malevolent but will desire some form of recognition, which implies sharing or controlling. There are pro-social, group-oriented cultural practices, but these rituals, objects, or events imply inclusion of individuals.

Evident in other species, we see mother and infant engaged in vocal or visual display and response. A mother nurturing her offspring displays an adapted behavior to help survival and longevity. Mothers teach and offspring learn how to respond to salient emotional signals, and this exchange is the basis of artistic culture. In her papers "Aesthetica Incunabula" and "Becoming *Homo Aestheticus*" art anthropologist Ellen Dissanayake argues that the early bond between mother and infant extending back to our prehistory is human ritual behavior in miniature and in nascent form. More than art-like signaling, says Dissanayake, these early interactions are on the one hand synchronized, consider babbling communication, and yet decoupled, where infants begin to think of past and future while engaged in the present.

From these cognitive-emotional foundations our cultural practices slowly evolved.

Culture and the Individual

Biologist Mark Pagel's *Wired for Culture* is an erudite examination of, to borrow from Richard Klein, the human career. While Pagel focuses on the universal aspects of culture, like knowledge, beliefs, and practices, much of the discussion hovers around the individual related to cooperation and the human tendency to form and adhere to small groups. Pagel places the blossoming of culture at around 80kya, by which time we not only learned from imitation but moreover began to innovate and re-engineer what we had learned.

We then passed that understanding to succeeding generations so that through an intellectual ratcheting-up effect symbolic artifacts like jewelry, paintings, and carvings began to appear widely. Beads and deliberately shaped objects hold and carry thoughts, values, and memories. Initially, bone and teeth pendants carried some abstract symbolism via nature or material culture to convey an attitude or belief. In time, as technology increased, these bone and teeth could be carved and shaped to express better the implicit values.

The bits of culture, from an idea to a technological feature, would function like a gene in terms of *transmission* and *reproduction* among individuals, says Pagel. In their classic study *Man the Hunter*, Richard Lee and Irven DeVore suggest that genes and culture co-evolve since sociality and ideology are as integral to human evolution as any physiological development.

In *Not by Genes Alone*, Richerson and Boyd say that culture is in part biological evidence of our adapted mind. And yet another part is flexible since some elements remain while others are discarded. As should be clear by now, our ancestry or genes do not totally determine behavior. If that were so, we'd have no culture and would survive and reproduce on instincts alone. The process of the adapted mind manipulating new elements creates a cultural environment that in turn selects for certain genes. If we and others approve of a behavior, it will spread and create its own population of ideas. As we will see in Chapter 4, some behaviors, for example artistic culture, are in part genetic. We are the only species fully utilizing our cultural capacity, suggest Richerson and Boyd, and so why we have so many large groups and mega societies with innumerable cooperating individuals.

Distinguishing themselves from evolutionary psychologists, Richerson and Boyd say culture is not evoked from an adapted mind but is transmitted through external means (i.e., not entirely by genes). They focus less on the innate universals of evolutionary psychology and more on environmental factors and group transmission. They see that individuals act on a population level. This might be true to some extent but does not account for a Copernicus,

Darwin, or even a Bill Gates, modern individuals, to name only a few, whose ideas have impacted culture as to have created different cultures.

Nonetheless, Richerson and Boyd's point seems to be: first a Darwin or Gates, and then the spread of ideas culturally. On some level cultural behavior is cognitive and not completely genetic. As there is complexity, diversity, and variation in the natural world, so there is too with culture. However, of course there is a genetic underpinning to what we call culture, or the genetics of human creativity, learning, and communication. There's an adaptive benefit *to behave* with others.

What then are the fitness benefits of culture? Culture permits individuals to learn and share with others knowledge at little cost. Richerson and Boyd do not see that genes entirely control or account for culture. Genes are sensitive to an environment that can help individuals in a group create culture with certain norms and expectations, whether of dress, symbolic markings, or values. For example, Richerson and Boyd see natural selection favoring cooperation in small groups, but the only way to explain our current social cooperation on the city scale is cultural evolution.

Cultural diversity demonstrates how on a population level any group adapted to the environment of its needs. Being among a group who share values and beliefs, more items and ideas can be produced and more people can be content. Culture springs from our adapted ability to have a self and to be aware of and cohere with others like us, as if melding the social and natural spheres. Roy Baumeister says in *The Cultural Animal* that culture is a "biological strategy" for interacting socially with others (8). In many ways cultural behavior is a resource.

Baumeister agrees that what we call culture, and often something idealized, as in attitudes about classical music and high art, did not rise wholesale. Rather, cooperative tendencies and consciousness, including self-awareness and theory of mind, linked together to tackle the real-life problems of food gathering, production, and consumption. Culture is driven by the biology of our brain and its innate needs and tendencies. This is why, for instance, some super-imposed cultures of the twentieth century have failed, like Soviet communism. We have survived and dominated not merely by intelligence but in how we have shared intelligence and the degree to which we agree with and accept packaged ideas. While there are other social animals, from wolves to apes, only we have built societies from knowledge.

A key claim of these thinkers is that our cultures, both the products and influencers of genes, are what count for our flourishing and survival. We are, says Pagel, in effect the "first species to throw off the yoke of its genes..." (4). Nevertheless, we are not inhabited by unchangeable robotic ideas, evidenced by our long and increasingly complex lineage and present survival, where cultural universals ripen in many different forms, flower fruit, and sometimes die.

Pagel suggests that we have an immune system for ideas: we do not rely completely on instincts but on consciousness that can be highly selective. The cultures we have created benefit us genetically. We have thrived and reproduced in communities. Our brain capacity is in part the result of our own inventions, a process of selective enhancement that continues. Our social inclinations and cooperative behaviors have served us in ways that were not available to other *Homo* species, like Neanderthals and Denisovans, with whom we shared part of our living history. They either died off or were perhaps overcome by us as we invaded and filled their cultural niches. Our abilities for symbolic thinking, for understanding beyond the literalness of an object, event, or spurious thought, and the subsequent social creations and structures that arose, are keys to our survival. For instance, Pagel presents the poignant image of Neanderthals on the edge of an ice cold Europe looking across to the warmth of Africa but not having the ingenuity to craft ships to get there.

The story of *Homo sapiens sapiens*, the wisest of the wise, is about those who could flexibly use the various modules and intelligences of the brain to solve complex problems. While Cecilia Heyes acknowledges that the mind consists of innate cognitive modules as per Cosmides and Tooby, she nevertheless sees development and non-cognitive mechanisms as important. What ultimately is responsible for any adaptation? It could be experience-based and phylogenetic (organism groups and behavioral) or systematic change over time and ontogenetic (individual organisms and cognitive).

Pagel's account of the crossing of the Beringia and of the so-called Lapita people inhabiting the Pacific islands by navigating thousands of miles of sea 6kya using stars as guides is breathtaking. Other *Homo* species made no spectacular migrations and simply adapted to their existing environments, apparently incapable of understanding, for example, that leftover bones from a hunt could be put to other uses, whether technological or ornamental. We, on the other hand, developed cultures to help us disperse across the world and spread ideas within and between our many groups. Without committing to a number, we know that there were many other species related to us, from the Ardipithecus about 4.5mya, the Australopithecines, and the Paranthropus with several species in each, and then to the early genus of *Homo*, which includes *habilis*, *rudolfensis*, and *ergaster*.

Why did we survive while they did not? Why are we so different? And yet why do we share behaviors with great apes? Building off Hatfield discussed in Chapter 1, our social learning is both continuous with but beyond that seen in apes. Social learning is a combination of behavior repeated for others and imitation of other behaviors. Learning involves mechanical manipulation and cultural adhesion. Even chimpanzees are capable of social learning with tool use that goes beyond imitation. But we possess, in line with my argument, what cognitive scientists Visalberghi and Fragaszy might call the blending of "individual experience" and "appropriate behavior" (264–65).

The individual can express herself, at times challenging the status quo so as to instigate change, at times following the status quo and simply holding new representations in mind for later use. We don't just socialize and learn, but we act with meaning. Our behaviors go beyond literal imitation. We understand the historical context and representational implications of behavior itself. Moreover, others grasp the significance of our meaningful actions and creations and respond similarly. Importantly, we can act and respond to actions purely through symbolic forms. The moral dimension is that we use artistic culture to influence the behavior of individuals and even large groups.

The Question of Cultural Transmission in Apes

In human evolution widespread tool use came via social learning as well as imitation, both of which include technical and social intelligence. In "Emulation, Imitation," Andrew Whiten et al. say chimpanzees imitate, but conservatively, and with more control than human children. Intelligence, as previously defined, is a problem solving function that involves repeated failure, success, and insight ultimately in terms of an individual contributing to the group. In other words, one must first understand that there is a problem to be solved, or an issue to be addressed, or a cause to be effected.

In his paper "Cultural Transmission," Michael Tomasello says that with free-ranging chimpanzees in equatorial Africa we see different behaviors from group to group in food consumption, tool use, and communication. Some differences are not ephemeral but generational. There is a question, however, about whether or not these differences are genetically inherited or even cultural. For example, close groups might show more differences than those much farther away. Some of these differences might arise from individual learning affected by the geographical and conspecific environments. The bottom line is that if there is no social learning there is no cultural transmission.

Tomasello says that cultural transmission consists in the passing on of behavior, either by learning or imitation. Across Africa, east to west, there are different chimpanzee practices. In the west there is termite fishing, termite mound penetration, and nut cracking. In the east there is termite fishing, ant dipping, leaf sponging, and ant fishing. Sylvan Soriano et al. make a similar claim in terms of Middle Stone Age technology across different locations in South Africa. While there is tool making culture, the practices are not homogeneous.

As Tomasello says in his paper, these differences are not necessarily based on geographical variations. With stimulus enhancement, for example, a chimpanzee sees a conspecific using a rock to crack nuts, so he picks up a rock and by accident cracks a nut. This action is not cultural transmission since the imitating chimpanzee does not quite understand or mentally represent the problem. There is no real attempt to reproduce the meaning of the behavior. Tool use learning moves from adults to children. A two year old chimpanzee observing its mother could take years to comprehend, emulate, and learn how best to fish for termites. Goodall relates how she once retrieved a stick used for termite fishing and unsuccessfully tried to use it.

But individual and social learning by chimpanzees is not truly reproducing behavior and therefore not cultural transmission, Tomasello claims. The apes do not quite understand the ideas involved in what they see and learn. Chimpanzees mimic a behavior that has no function. Even when a chimpanzee imitates, there seems to be a lack of understanding the applicability of what is being imitated. For example, potato washing by Japanese macaques took almost four years to spread, and at that only to half the population.

William McGrew, on the other hand, does see a cumulative progression in this behavior by Japanese macaques of the Koshima Island. What began as potato washing became, over time, a way to season the potatoes with salt as well. In turn, this practice gave rise to similar cleaning of wheat. Ultimately, the Koshima monkeys, with prolonged contact in seawater, now fish. Sarah Marshall-Pescini and Andrew Whiten do not discount chimpanzee intelligence for simple cumulative culture but doubt there is anything close to a ratchet effect here. Nevertheless, Whiten in "Ape Behavior" says that stone handling among the Japanese macaques seems to be cultural transmission primarily because there seems to be no *function* for the behavior.

Tomasello's point is that while we see learning and imitation in chimpanzees regarding communication, tool use, and other practices, these emulations and manipulations do not truly constitute the reproduction of behavior, which probably accounts for the lack of ratcheting up progress generationally. Indeed, there was human intervention with the Koshima monkeys, where people provisioned them near the beach. Is it fair to say we do not know whether or not there has been any ratcheting in ape culture, since we do not know the earliest forms? Whiten points out, in answer, that chimpanzees of Goualougo hunt for underground termites, first, with a stiff stick to dig, and then, with a multi-edged soft tool to collect many of the insects at once. This is not behavior that could have sprung from one step.

Jane Goodall cites how eating traditions and cultures change among chimpanzees. One young chimpanzee experimented with mango eating and over time more were seen to eat the fruit. Prior to this incident in 1965, Goodall says that she saw no chimpanzee express interest in mango, and yet these animals have quite a varied diet and keen interest in fruit. Meantime, by 1982 Goodall saw the first mango eaten by a five year old. Maybe this simply confirms Tomasello, if it is only imitation. If it is an instance of a chimpanzee crafting a new behavior, as in food experimentation regarding taste, then it could be labeled cultural.

Similarly, in chapter eighteen of *The Chimpanzees of Gombe*, Goodall spends quite a bit of time on object manipulation, which can go beyond mere imitation to behavioral understanding. Use of sticks for attack or intimidation, twigs for insect probes or dental cleaning, leaves for sponging food or hygiene, and stones for intimidation or nut cracking can vary from community to community. There is developmental variation, suggesting that the chimpanzees on a cognitive level understand and creatively transfer behavior. Effective tool use techniques, which can arise from one individual, can then pass over generations. Goodall suggests that both change in and altered copying of behavior constitute cultural transmission.

Culture, Social Learning, and Adaptive Benefit

Our species went beyond stimulus enhancement, repeating in different environments what one would do anyway, to social learning, which is deliberate awareness to design in order to improve a behavior or tool. We have *invention* by intention and not by chance. There might be a genetic basis for constructive designs, since we find similar artifacts in widely different places. In terms of cultural evolution, anthropologists Joseph and Natalie Heinrich say we tend to imitate those who have succeeded and possess prestige, based on inborn survival mechanisms. Since we don't know why the person is skilled or successful, we tend to copy traits. At bottom we are intellectually able to process and comprehend the significance of another's behavior. With a lack of concrete information, we gather what we can and adaptively conform to the mass if it realistically or seemingly benefits us.

There are human artifacts like stone tools that are very old, but there is almost no improvement in those tools for over one million years. While true, this statement can be deceiving. Steven Mithen dates the earliest stone tools, flaked quartz, to almost 3mya. Early tools gradually and dramatically improved. There are heavy duty and light duty tools around 1.5–2mya, hand axes and Levallois flakes around 1.5mya-250kya, blade technology and flint slivers before 100kya. Michael Balter has made the important point that tool making could be a proxy behavior for early symbolism. Balter goes on to say that brain scans of skilled, current stone knappers reveal, for the manufacture of early tools, activity in visual and motor areas, while for the fabrication of later tools, activity in language areas as well. Tool manufacture is not just imitative but behavioral.

If ancient human species had a very rich, inner cognitive life, it would probably be reflected in the artifacts they produced and left, some say. This neglects the observations made by Darwin in that dominant forms prevail. If there was not enough modification, there would have been extinction, but early cultural forms helped us, rather than Neanderthals, to survive. After variability there will be some constancy, and that seems to be the case with the stone tools that were used for digging, chopping, and butchering.

Yet the same tool design persisted. Perhaps the early hunter-gatherers needed not to improve anything over long periods since they had not yet begun to form larger groups and then tribes. Crucially, in spite of severe climate change, *Homo ergaster* and *Homo erectus* survived with these tools. Early people could have had a richer individual cognitive life that is not necessarily reflected in their quite serviceable tools. Since Pagel's claim is that we improved on learning, there is no way for us to discount the possibility that some random, discrete individuals had nascent, rich cognitive lives that had yet to see efflorescence and copying. Cultural eruptions were embryonic and had yet to find the right catalyst, probably when our brains began to make connections in finding new and different uses for existing products.

We begin to see in caves of the Western Cape Province, South Africa, evidence of cultural (not just technological) artifacts dated to 160–75kya. European cave paintings at Ardèche and Chauvet-Pont d'Arc, France, date to around 36kya and Lascaux at about 18kya. Genevieve von Petzinger and April Nowell say evidence from Africa and the Middle East put symbolic expressive behavior back to at least 100kya, meaning that Chauvet might not be an anomaly. AMHs, by this accounting, would already have been involved in complex, symbolic behavior well before Chauvet. The geometric shapes found in the cave paintings, von Petzinger insists, such as dots, line patterns, and cruciforms, are manifestations of older techniques.

Of course there is red ochre dated to around 360kya, but no one is quite sure what that might have been used for. With ochre manufacturing at an early date, it is fair to assume body decoration indicates that symbolic artistic culture is much older than previously thought. Archaeologists Christopher Henshilwood et al. show that there are actual ochre producing toolkits dated at 100kya, which suggests early people used it for some purpose. Chris Knight et al. claim red ochre was used by early women to falsely signal menstruation.

The capability for cultural manifestations was latent but only flowered once we adopted social learning. Cultural genesis and development come from mind and not necessarily from genes, though genes are in play with neurons and neural connections. For example, our predecessors roamed the globe and adapted physically, genetically, to the new environments. But it was, says Pagel, culture that initially propelled them and subsequently sustained them to form complex social divisions of labor and societies.

There is of course an advantage to different cultural groups in terms of variation and competition. There is also the passing down of rights and property, an advanced development, and hence why, Pagel suggests, there are many different cultures. "Cultures restrict the flow of genes," he says (54). In spite of one *Homo sapiens* species we all look ethnically different by virtue of our ancestors' predilections for selecting certain features. That selection is mostly cultural. Indeed, there are cues to which we respond positively when identifying who is part of our group and negatively when noting who is outside of the group.

While Pagel stresses the importance of the group in terms of culture, he does not minimize self-interest, saying that selection favored in us a mentality to form cohesive groups. Regarding cultural evolution and social learning, the "variety" of individual skills and talents count most, according to Pagel (100). Culture is a sorting process so that, for example, someone has the ability to make a musical instrument, and then someone else has the ability to play it well. Pagel makes a sustained argument for diverse cultural groups, but this development can only come to pass through distinct individuals. People with similar beliefs or inclinations merge and develop cultural practices. Oddly, individual differences have little to do with overall human survival. Yet natural selection would then have eliminated our differences. There is an evolutionary bias for a diversity of personality types.

Likewise, Pagel says that the arts and religion are *cultural enhancers* or emotional motivators related to behavior. He seems satisfied with the simpler notions that the arts transmit ideas and that religion helps explain occurrences. Is this a bit perplexing? Pagel is arguing that we survived and thrived because of culture, but he does not quite come out and argue for an adaptive function in the arts and lumps arts together without distinguishing one from another. Or maybe he does argue for an adaptive function. Even if beliefs are wrong, false, or incorrect they might, nonetheless, help a group survive. Indeed, Scott Atran says that fMRI studies show theory of mind brain regions are involved in discussions of God, and that counter intuitive beliefs are easily accepted since we are socially and culturally motivated toward our own group.

Even without religion we would be much the same, i.e., self-interested and morally corruptible. Simply put, says Pagel, we have concocted religion to offer ourselves "courage and hope" and to *coordinate* and *unite* groups over other groups (159). As natural selection pits genes against genes, so religion *induces* emotions shared in a group opposed to another group in "cultural relatedness" (165). Conflicts can render an opportunity for a group to produce what it believes is a moral outcome. There are many distinct cultural groups with different beliefs though all with one common denominator, the need, apparently, to own beliefs.

Pagel notes that reciprocal altruism by virtue of its mental complexity exists only among human beings. Altruism is always on shaky ground as game theory has demonstrated. In the words of political scientist Robert Axelrod, notes Pagel, there is "the shadow of the future" or the possibility that one party will look ahead to extraordinary gains and so default on any agreement (191). Nevertheless, we all seem to be programmed not only to be fair but to be generous, since we expect to be so treated. Paradoxically, fairness is rooted in self-interest. Social interactions depend on theory of mind, which in turn depends on a brain very different from that of a chimpanzee or our own prehistoric ancestors. We are able to detect deception quite well. In spite of all this talk about culture and groups, individual motives, needs, and desires count. We especially see individuality expressed in the creation or use of artistic culture.

As noted in Chapter 1, the rapid development of our brain in quality and not just quantity helps explain our survival over all other hominin species. Many of our cooperative behaviors are a result of language and the ability to navigate multi-party transactions. *Homo ergaster* had very rudimentary speech followed by more advanced speech in late *Homo erectus*. Pagel insists that only our species produced language, at most 200kya, because of "social complexity" and as a "trait for promoting cooperation…" (279, 281). Recalling Dunbar's gossip theory, Pagel notes that we use language, across the globe in about 7,000 current forms, not just to speak but "principally to talk about each other…" (294). Language sounds are reflected in key resonances and word lengths universal for thousands of years. There is a similar foundation to all languages in that we can translate ideas from one to another. We might have an innate desire to learn language just so our ideas can be understood. Before children become fully proficient linguistically, they make art to express their ideas.

In his discussion of free will and the function of consciousness as a cultural operator, Pagel says that our brains work out patterns ahead of schedule so that the subconscious might already know what to do in certain instances. This is relevant since our culturally-hungry brains are always in operation mode, as if on a sixth sense. Relatedly, an older part of our brain affectively responds instinctually to highly charged moral situations – do no harm. As neurobiologist Kathleen Rita Gibson would suggest, instincts and intelligence are not unrelated mental abilities, in that both are responses inherited and related to environmental or other cues. Clearly, social sensitivity has become over time a key cultural ingredient. Interestingly, though, Pagel suggests that we do not know ourselves because of "introspection" but because of "observing our own behaviors…" (327). Often we don't know how we'd react in a hypothetical situation. Artistic creations supply us with simulation, and so moral dimensions inhere in artistic culture.

Pagel goes on to say that consciousness is little more than an after effect of a highly active brain organizing input. There might be something illusory about what we label "I." Yet Pagel does not seem to be hinting at cultural determinism or the standard social science model in learning. Rather, he seems to suggest that in order for us to imitate and improve upon behaviors, the cultural tools that have preserved our species, our brains need to catalogue seamlessly various strands of information in advance of our conscious processing of information. As I have been arguing, these choices, whether of creation or imitation, convey moral attitudes of approval or disapproval.

Finally, coming back to social learning and our need to connect in clusters, Pagel notes that in spite of our near obscurity in large cities Milgram's notion of six degrees of separation is valid. We are interconnected by people who know others who know us. The point about being human is that *social viscosity*, where we form small groups and stay close by "social rules," has not changed much over our long evolutionary history (365, 367). Group living, however, does not discount individual differences or self-interested behavior, even if this behavior is creative and challenges the group thinking in some radical manner.

Culture and the Group

Evolutionary psychologist Alex Mesoudi's *Cultural Evolution* takes a different approach and leans more toward group selection. But even genetic hardliners and those who emphasize individual character over the influence of situation can be persuaded. According to early cultural anthropologist E.B. Tylor, culture is "'that complex whole which includes knowledge, beliefs, art, law, morals, custom, and any other capabilities and habits acquired…'" (189). Culture is subject to Darwin's model of variation, competition, and inheritance, for it benefits the group. While some might rightly turn aspects of Darwin's ideas toward group selection, he of course fundamentally focuses on the individual. Although he was an original thinker, Darwin was a product of his time and influenced by Lamarck (inheritance of acquired characteristics), who does help us understand cultural evolution.

Mesoudi says that "individual learning and genes cannot fully explain human behavioral variation…" (12). Size and fluidity of population figure into this equation. So there is reliance on cultural explanations or what Lamarck and even Darwin at later times would call use and disuse of traits in terms of inheritance, as opposed to genetic transfer. Mesoudi points to the well-known example of the holistic outlook of East Asians compared with the analytic outlook of Westerners. Granted, that is very general.

Nicholas Rule references neural studies that demonstrate how a Westerner is more object-focused whereas an Easterner is context-oriented. Cultural influence can affect how one sees people, the most prominent figures in our field of vision. Even theory of mind, says Rule, has been shown to be culturally dependent so that Asians read Asian eyes better and Caucasians read Caucasian eyes better. At what point does a cultural predilection become genetic?

Mesoudi says that B.F. Skinner, by focusing on individual learning and conditioning ignored culture. Yet he is also critical of Tooby and Cosmides who lean more toward genes with their notion of an *evoked culture*. Echoing Peter Richerson and Robert Boyd on whom he relies, the upshot is that "genes alone cannot explain human behavioral variation" (13). Some might debate this conclusion, evidencing variation as one of the pillars of Darwinian thought and how individual difference helps propagate the species by ensuring that some offspring will be better adapted with more beneficial mutations than others.

Those like Mesoudi see genes as only responsible for certain potentials, i.e., learning itself, but not for content, i.e., values or beliefs. This seems to ignore basic universals including parental care and fairness in reciprocity. Robert Turner and Charles Whitehead, in "How Collective Representations Can Change the Structure of the Brain," make an argument similar to Mesoudi's. For example, one is not born with a talent or even an inclination to be a musician. One becomes a musician by constant exposure and practice.

By this line of reasoning, anyone could become a great musician. While environment or culture affects an individual's interest, where do personal drive, determination, or discipline come from? Those three Ds are in great part genetically based, as psychologists like Jerome Kagan and Elaine Aron have proved. Determinism does not strictly equate to behavioral outcome, and that is where Mesoudi is somewhat correct.

However, cultural transmission and social learning alone do not account for individual variation. Cannot the learning potential of individuals vary, and is that potential not genetic? One can argue that we have everything Mozart left behind because of the guided influence of his father and the prestigious musical culture of Austria. But then, one could argue that Mozart was a genius constituted uniquely from his family's genes scrambled. Certainly a cultural environment can help the expression of one's genetic talents. Others, too, who look for a biological explanation of individual morality might quibble with the assertion that genes are not responsible for content. As Schopenhauer argues, it matters not whether one gambles for bits or nations. What matters is how honestly or not one behaves, i.e., how genetic self can be definitive in sculpting behavioral outcomes. Genes make a difference.

In the early sections of the book, Mesoudi understandably keeps steering away from individual genes in terms of learning and favors culture. No one would completely oppose this thinking. Cultural learning is easier, more adaptive, and works faster, but individuals have innate dispositions that can affect learning and outcomes. See, for example, Rimfeld et al. where genes account for academic ability. Regarding learning, the human species alone has an advanced, cumulative, and highly influential culture, but how do we explain the many discrete individuals, not groups, who resist some culture?

This is not to take away from Mesoudi's points or argument. Mesoudi goes on to argue convincingly about Darwin's model of variation, competition, and inheritance in terms of cultural transmission. Of course in this model there is acknowledgment of individuality, i.e., variation. Without completely eliminating the influence of environment, Thomas Bouchard has concluded that there seems to be a biological imperative that falls, rather, to the effect of inherited and distinctly personal genes on one's psychological constitution. There is also variation within and between groups, which helps establish cultures. A massive study on twins by Tinca Polderman et al. concludes that multiple genes combine to pressure an overall sum effect on a trait. Genes contribute to traits as much as environment. Maybe more, since one's genetic temperament will guide one's decisions and behavior: we often choose which group to belong to.

Different groups, extending back to our prehistory, are different cultures in terms of languages, religious practices, and social customs. More specifically, as per Darwin, the competition will occur between like species, since both are trying to secure a certain cultural foothold and vie for the same resources. Just as there are distinct individuals and groups, there will be competition, and in cultural competition there will be ideas against other ideas, skills against other skills, evident to this day among fractious religions and countries.

In a Darwinian manner, some aspects of culture can become extinct. As Darwin speaks about inheritance, we can see from observation and documented history that values, beliefs, ideas, knowledge, and customs are passed on within families and within groups. Some ideas and beliefs are modified considerably, some not at all. Some die. While there is in cultural transmission, on the one hand among human beings, very close imitation, there is also, on the other hand, descent with modification. We have progressed and flourished precisely because whatever we learn, whether in manufacturing a product or in generating an idea, is improved over successive generations.

In addition to writing on a Darwinian application of cultural evolution, Mesoudi manages to weave into his explanation a brief history of early cultural evolutionary theories, including Herbert Spencer and his insistence on *progress* to Lewis Henry Morgan and his European-American centrism in terms of classifying races. Whether products of their time or simply ignorant, these men incorrectly state that some cultures are more evolved than others. In terms of object manipulation, Jane Goodall cites two separate examples of two different chimpanzees who, after being touched by a stranger, wiped the area with leaves. These reactions are performed for whose benefit, the group or the individual? Goodall reports that typically, strangers are not liked, which would account for cultural group selection.

How does one define and qualify one *culture* against another or aspects of any particular culture? An early, small group, the Eighteenth century British Quakers, spoke out against the slave trade long before it was abolished. Were they counter culture even if theirs is the position that ultimately succeeded? Sticky business, what we call culture and risky to speak about it in general terms. The vastness of humankind and the range of populations have both complicated what we call culture, in spite of some basic cross-cultural emotions.

While Darwin might not have *meant* the word to carry the force with which it strikes many readers today, he does after all use the word *savages* frequently in *Descent*. In spite of his genius of observation and single-mindedness was Darwin a product of his class and culture? He was fundamentally against slavery but profited from his family's factory workers in Wedgwood china. The key question is whether one can escape the influence of culture. Since some cultural influences are harmful, can individual genes offer an antidote? Or, in terms of character, at what point is circumstance self?

There is no question that we have great capacities for learning, copying, and for innovating behavior. The big question concerns how much of what we are or become is from circumstance, the environment of parents, peers, and culture, and how much is driven by our individual DNA. Generally in line with the

Polderman study, most experts split the influence of genes/environment 50/50. From where I stand, that means half of what you are or become is genetic.

The above paragraphs simply prove Mesoudi's claim about Darwinian variation, competition, and inheritance of culture. Cultural evolution is not ladder-like and not progressive, as Spencer argued. Societies do not progress up a series of steps to a pinnacle, since there is no top. Rather, there are variations within a population that through natural selection cause change over time. Often, one culture borrows from an earlier version of itself or from another culture. And this brings us, Mesoudi sees, to the crucial question as to whether or not the transmission of culture is *particulate*. Turner and Whitehead cover this question as well. Biological inheritance is on the micro level and is not, on the surface, a blending of traits. An individual gets only one version of any gene, not a blend of it, and Mesoudi provides examples, including the color of eyes or fur. Darwin, unfamiliar with genetics, favored the blending model, but taken to its logical conclusion, at some point all traits and characteristics in a family would be blended out. Not so with Mendelian genetics.

According to Mesoudi, cultural traits can blend, with the example of language. Or not, since on a neural level aspects of culture, like sounds in a language, can be discrete. Lamarck, Darwin's predecessor, espoused use/disuse and the inheritance of acquired characteristics. Mesoudi says that cultural evolution is Lamarckian. We do not acquire neural activity from others but copy behavior and then modify it before passing it on. In a nutshell, according to Mesoudi, cultural evolution is Darwinian, just not neo-Darwinian, the latter espousing the mathematical and genetic models of the evolutionary synthesis in the 1920s/30s by Fisher, Haldane, and Wright.

So part of this discussion is on microevolution, or experimental genetics, and macroevolution, or the naturalists. Microevolution deals with the transference of particulate, genetic matter in a defined population. Macroevolution deals with cultural change across populations. For instance, according to R.A. Fisher, female choice increases the genes that increase the genes for a trait as well as the genes for the choice, as in a pea hen preferring the peacock's tail. The pea hen's genes for preferences are transferred to both her male and female offspring. Rather than being blended out, the trait is reinforced genetically. In terms of cultural evolution, an equation like this can exponentially increase a trait to become exaggerated and spread. Consider varying cultural attitudes toward sexes and genders, in spite of a general valuation of family.

Concerning descent with modification in culture, what came first? Darwin's early critics, e.g., Fleeming Jenkin and Lord Kelvin, complained of a similar problem. How did natural selection start? Mesoudi goes on to offer an answer, provided in part by the phylogenetic tree (evolutionary branching). Societies, while culturally distinct, nevertheless share traits, all of which go back to some common ancestor. So while cultural transmission, some argue, is a messy horizontal blending and does not therefore lend itself to a phylogeny, analyses have nevertheless been done.

Mesoudi cites with authority Judith Rich Harris since the model promotes horizontal (group) and not vertical (parental) learning and transmission of information. Meantime Jerome Kagan questions Harris' research since his emphasis is on discrete amygdala activity and individual temperament. But Mesoudi's argument is understandable since learning and cultural evolution may come down to a question of scale. The parents are but a small part of the larger and more informed group which is well-stocked with prestige models of skilled workers and expert teachers.

Philosopher Chris Buskes also extends Darwinian thought to explain how culture evolves. There is variation-difference, selection of fitness-enhancing practices, and replication where ideas are transmitted across generations. Cultural evolution is cumulative selection, predominantly human. Culture evolves but is not guided or designed since elements can disappear and then mutated ideas suddenly appear.

From Nonhuman Culture to Human Artistic Culture

Is there culture in other species? Concerning chimpanzees, Jane Goodall answers affirmatively since apes of various locales demonstrate different preferences for hunting and tool use. Mesoudi looks at rats, guppies, rhesus monkeys, female quails, songbirds, octopuses, and honeybees, who engage in social or non-genetic adaptive learning. Human beings, however, move from social learning to cultural traditions in and among groups. Human culture is Darwinian since it is cumulative, nearly exact transmission yet descent with modification.

Boyd, Richerson, and Henrich insist that many human adaptations to ecological problems are cultural and not genetic. Since we do not have, contradicting Cosmides and Tooby, dedicated domain specific intelligence modules, we can improvise intelligence flexibly to different environments. The authors cite, e.g., the Inuit of the deeply cold climates whose adaptations result from accumulated cultural information and learning.

Without doubt the study of life must include the study of culture, and any study of culture must include biological sciences. Beliefs, values, and practices are part of more advanced life forms and affect how living organisms function individually and among each other.

As Richard Alexander might say, we've developed our own selective costs and benefits through sociality and our *runaway intellect*. Our minds evolved to solve interpersonal and group challenges. As we solved problems, related issues arose, but we solved more and so were able to solve yet more because of our greater capacities. Adaptations like imagination and creativity provide a social function between individuals cooperating or competing in a group and cooperating or competing between groups. Cooperation and competition could be for attention, e.g., reproductive advantage, or status, e.g., resource advantage. As a social-mental function the mechanism could be art.

Inspired by Frans de Waal's *Age of Empathy*, and pulling together my argument to lead into our final chapter, let me offer the following evolutionary flow chart that attempts to consolidate cultural-moral behaviors and artistic display mechanisms:

> Individual action > mimicry > synchronized group activity > facial or gestural mirroring > body contact > shared emotions > empathy > emulation > new artistic behavior.

Now we see artistic culture as highly sophisticated, but in an ancestral environment any novelty that provided an advantage would have sufficed in a cost-benefit analysis. From prehistoric time we have evolved ways to influence another's behavior, and that has been the environment in which our adaptations have evolved. At some point individuals were able to create what-if scenarios, and these attracted attention, made predictions, and offered solutions. Individuality was prized and rewarded. From novel, creative social communication human arts, as we now know them, arose.

CHAPTER 4

Art and Aesthetics as Moral Cognition

The purpose of this book is to demonstrate how human evolution, especially in terms of cognition, gave rise to cultural practices. Culture in turn contributed to our art making tendencies. Because art behavior and artistic culture either approve or disapprove the values and beliefs of others, I've labeled these practices as moral.

Early interpretations of cave art, says Margaret Conkey, viewed the works as some form of magic. Initial evaluations were based off a Eurocentric and Roman Catholic history. From a Eurocentric tradition of art history, we see artistic work as almost separate from everyday life. While art can be, and is according to my argument, value-laden, it is in some way roped off from the diurnal in its own sphere. But this is not so with prehistoric art. What we term prehistoric is really, says Conkey, a material culture of hunters and gatherers. There is no separation of function and symbolism. A pendant shaped as an animal *is* the animal. One does not simply make an art object in a hunter-gatherer society. The making of the object is participation in a social *and* symbolic circle and so carries moral weight.

Paleoanthropologist Robert Bednarik, furthermore, says that outside of Europe Pleistocene art is ignored, especially that of Asia. He reminds us that art was not invented in Europe. Nineteenth century Europeans could not accept that prehistoric people could be so competent, and so they politely disapproved of what they found. Bednarik, for example, identifies petroglyph rock art cuttings in India from the Lower Paleolithic and made with Oldowan tools.

In the preceding chapters we see that there is an arrangement of adaptations that gave rise to how and why we make artistic culture. Prehistorian Jean Clottes suggests that some markings, like the groupings of red dots in Chauvet, appear to be signs or symbols. Cognitive? Art in remote caves by adolescent boys, confirmed by the handprints found, would be part of the young male syndrome. Sexual? Boys will take risks to enhance their reputation competitively. Status? Art can be a means for outwitting others. Social? There can be multiple, interlocking adaptations to explain artistic culture.

Just as our brains enlarged for a number of reasons, such as selective pressures in a social environment, so too did that brain manage to develop tools for butchering. Eventually, different modules of the brain connected with longer neurons, so that physical objects like tools could be associated with beliefs and practices, giving rise to artistic culture. In this way a small stone could be a tool

 | DOI 10.1163/9789004319486_006

or a bodily pendant, and this pendant could be a charm, status symbol, or a group marker.

Added to these likely scenarios is Darwin's theory of sexual selection where artistic culture might have been a response to reproductive pressures, much as the peacock's tail or male bowerbird's bower. Stephen Davies, however, says the bowerbird's creation is an extended phenotype (genetic impacts on the environment for reproduction or survival) and not simply an artifact. Ornithologist Richard Prum argues that artistic display co-evolved with the aesthetic response to judge the display and its own evaluation. An aesthetic experience, both emotional and cognitive, evolved in response to a continuity of complex biodiversity.

We and other animals either approve or disapprove of display signals for mating or other fitness enhancing resources. From any response to basic forms of display an emotional-cognitive feedback enables an organism to imagine elevated forms of the same display. Over generations value judgments about that display evolved. Prum offers bird songs as examples. What some female birds now hear is actually the evolved product of selective decisions made by their female and male predecessors over many generations. Choices of approval or disapproval carry moral dimensions in the human sphere.

Artistic Culture and Moral Behavior

The question is not what is art but why make art. As I have been arguing, by virtue of our sociality and extraordinary cognition, much of what we call the arts springs from a moral core. This moral staple is grounded in natural feelings to communicate, to share emotions, and to challenge others. Nicholas Humphrey has said that human intelligence evolved in order to think about other people, and this is a moral attitude. Concern about another's behavior in terms of approval, honest and fair, or disapproval, dishonest and unfair, is social and so in turn gave rise to artistic culture or stories, religion, and philosophical ideas. Paul Bloom, too, says that fundamental to what we call arts are questions of right/wrong and good/bad related to human behavior.

Art is individual representation socially communicated, an arbitrary language we nevertheless accept and share. In *Consciousness Regained*, Humphrey suggests we create culture to stimulate certain experiences and feelings so as to inform and to lead outwardly our inner subjective self. We broaden our ability to understand the social habits of others as a type of identification by simulation. Cognitive psychologists Samuel Moulton and Stephen Kosslyn assert that mental imagery is simulation for as-if scenarios. What is the function of imagery? Because mental imagery is simulation, it permits us to predict from experience and so to bring outside reality into our minds.

What propels us to use our imaginations to create, to immerse ourselves in the arts, to spend vast amounts of money to read thick books, watch long movies, and listen to intricate music? The answers are complex, but in part deal with our evolved brain neurobiology that keeps us individually unique but yet desirous of mixed company and attention. Artistic culture is part of our social strategy for inclusion, separation, and distinction.

My thrust has been to emphasize the so-called moral components of the evolution of art culture. An individual or group can exercise degrees of care and concern or protest and harm with artistic culture. Some thinkers, Geoffrey Miller and Denis Dutton to name two, have followed Darwin's ideas to place all our arts in the arena of sexual selection, a self/gene-centered behavior. While I do agree with sexual selection, for human beings it is more nuanced since we are endowed with a complex social brain. Hence my emphasis on what I broadly call moral behaviors since our conduct invariably impacts others.

In *The Ancestress Hypothesis*, following Richard Alexander and Ellen Dissanayake, anthropologist Kathryn Coe has de-emphasized male competition and sexual behavior to focus on long-term maternal care in establishing cooperative traditions. A human male becomes a musician for more reasons than simply to attract women. Besides, in most species male bowerbirds make new bowers each year, mainly because of habitat changes. The point is that human art culture tends not to be temporary or disposable. Rather we place high value even on simple yet aesthetically pleasing artifacts and pass them down in families or among friends. That activity carries moral implications.

Artistic culture can join people, as in ritual. Artistic culture can separate people, for different groups have different traditions. Just as we say that the best writers have distinct voices, artistic culture is a means for the individual to call attention to herself. That behavior is social, or anti-social, and therefore is tinged with moralistic concerns. For example, does the created work uplift and carry the current norms or smash them? Why do we have symbolic forms in the first place? In *The Symbolic Species* neuroanthropologist Terrence Deacon says that our cognitive ability counts far more than our physiology, and that we spend most of our time thinking about past, present, and future. Without question this thinking is in terms of our own behavior among others. We ponder and agonize over what we did, could have done, or should have done. We can use artistic culture, both as a maker or participant, to correct, amend, or elaborate our feelings about our actions.

We are obsessed with what others are feeling and thinking, to the point where we try to imagine what an artist was imagining when she painted.

Human beings are drawn to the aesthetic, which is both cognitive and emotional, so somewhere in our evolution the two became twined. Brain areas don't simply light up in an aesthetic experience, comparable to a chimpanzee seeing ripe fruit. Instead, what we call aesthetic is part of our intellectual faculty for symbolic communication. In this way we can channel our emotions, either as an artist or viewer, in ways unlike our ape cousins. As Francis Steen would suggest, any cognitive account of aesthetics points to the experience as adaptive since it is implicated in "self-construction" (57). We are not automatons. We are not instinct-driven. We take pride in our individual identity, which we manufacture, and part of our identity derives from adherence to artistic culture we choose, or not.

In other words, the drive for the aesthetic is not the same as the drives for other resources since the aesthetic experience itself is somewhat immaterial and more emotional and cognitive, a means for recharging the brain individually or collectively. Art creation or thinking is adaptive since in its representations it points the way to new patterns that can be useful in addressing behavioral, environmental, social, or practical problems. While now we might not see those practical concerns addressed in cave paintings, we can appreciate their emotional appeal. Art is pretense, and pretense is basic simulation to create options for new orders. Artistic culture helps us see differently, and we not only require but also enjoy imaginative activity.

We have an innate bias for creative thinking to solve problems and overcome obstacles, and this mental activity is social. In terms of our nearest primate relatives, Goodall says that chimpanzee grooming is to support or improve relationships. Grooming is a social and mechanical function. There are recorded instances of leaf grooming to initiate another's attention, but this can happen solitarily, as if the chimpanzee is *doodling*, says Goodall. This is not to say apes are artists, but merely to demonstrate how culture is a morally social behavior, especially if it cuts across the grain of the status quo in an attempt to alter culture.

Köhler quite often saw chimpanzees deliberately holding, carrying, and displaying on their bodies any type of string, rope, or rags. One even wore a metal chain around her neck. More recently, Bednarik talks about chimpanzee self-awareness via self-decoration. He cites a number of sources and points out that self-decoration occurs in big brained species who have large social networks. The brain's feature detectors, neurobiologist Kathleen Rita Gibson tells us, prefer singularity in line or color, clearly evidenced in these behaviors.

Further, neuronal networks like those found in primates are interconnected to detect for colors, shapes, movements, sizes, distances and positions at the same time. Why? We evolved those mechanisms in response to selection pressures related to social survival and reproduction. This is what makes artistic culture, as I have argued throughout, *moral*. Our brains have evolved adaptations, expressed creatively in visual and musical arts, which helped address and perhaps solve communal issues, personal dilemmas, and interpersonal problems.

While more neurons help in detection, connections are paramount, says Gibson. An organism does not just have the ability to detect one million different, single objects but to make multiple detections simultaneously. From this evolved adaptation to process much stimuli simultaneously, along with social pressures, arose our capacities to transform those neurobiological networks into material objects.

Art and Adaptation

Stephen Davies says that Hegel and art critic Arthur Danto find post-eighteenth century art a vestige, a feature without function. Art no longer has a purpose, in this view, and simply persists like the tail of a comet. In *History of Art*, Janson says that because we place high value on art we usually separate it from our routine lives. We put art in caves or museums. This is a Eurocentric and modern view. Before material culture eventually became high art it was functional. Janson does acknowledge that there are universal brain mechanisms shaped by culture and that there is an ethnographic prehistory to art.

My concern is less about, as Janson would say, the *ideal form* of art and more about its moral evolution. In the end, the question of adaptation everyone quibbles over might be simply the mind's cognitive ability to produce art behavior in response to other stimuli and needs, such as sociability and creativity. Yet any response or need is one of approval or disapproval, which carries moral import.

Philosopher W.P. Seeley says aesthetic cues have always been in nature. Neuroscience, to be discussed momentarily, is only good at explaining the how and why of our visual experience. Rather, Seeley suggests, we have knowledge that helps us structure and evaluate art perceptions. Consider how in our history we've always been watching, hunting, and responding to animals. Seeley does not indicate whether this knowledge comes from an adaptation or from other artistic culture. Salience or what is revealed from an object or event is vital, but how did we either adapt or learn to cue any salient feature?

Evolutionary psychologists John Tooby and Leda Cosmides ask the question in the title of one of their papers: Does beauty build adapted minds? Imaginative creations are found across cultures, so the pleasure we experience in art has persisted to satisfy a biological function. By eighteen months children enjoy inventive play, which requires perspective taking, a social function. Tooby and Cosmides say pretense is an important cognitive advance concerning representations to help differentiate between reality and the fictional world. Mental imagery is decoupled from hard reality so as to become useful in solving problems. Our aesthetic sensibilities are adaptations related to locating desirable locales or foods and attractive mates. As Tooby and Cosmides say, we evolved an appreciation for beauty because certain objects were advantageous to us and so demanded more mental and visual attention. When we encounter a visual cue, we are able to value-weigh it against our inborn and learned representations.

Art anthropologist Ellen Dissanayake says contemporary human beings cannot be polled about art preferences, as artists Komar and Melamid did, to uncover Pleistocene tastes. See too Richard Coss, who demonstrates our evolved perceptual biases for canopy trees and our predisposition for watery surfaces. We might like highly polished objects since we evolved around glistening lakes. Art behavior is a complex response and not simply a cue-reflex according to an adaptation.

Dissanayake says there are emotional and cognitive components that constitute an art response, melding concerns involving human relationships, sexuality, life and death, or sickness and health. These signaling elements themselves are not artistic culture. Art is not just an adaptive preference. We need to explain, Dissanayake says, why we have evolved a preference for aesthetic experience that heightens our cognitive emotions. Artistic behavior shapes components that make our cognitive and emotional responses. This shaping carries moral tones in that art can either uphold or challenge norms of behavior.

When I say cognitive emotions, I mean that cognition might support art as a byproduct. That stand does not eliminate emotions from the equation. Our instinctual and innate responses are connected to art behavior, so there is an evolutionary component to art. I suggest art is adaptive since it has evolved from pressures to approve/disapprove, both as social survival mechanisms.

Art is implicated in gene/culture co-evolution since physical representations are mental information stored externally. In turn, viewing artistic representations affects one's cognition which can have an impact on gene expression and behavior. There is much speculation about prehistoric art, but undoubtedly it served a number of functions, including personal display, status classification, group demarcation, and ritual purposes. Importantly, all of these hold symbolic value and hence the ancient cognitive connections to human culture. Archaeologists Steven Kuhn and Mary Stiner say that creativity is "certainly a cognitive phenomenon…" (143). As outlined from Chapter 1, creativity happened early in our prehistory. The same artistic culture in the form of tools persisted for many thousands of years since only changes in resource retrieval would have prompted a slight change which would have had a ripple effect on cognition and behavior.

The hand axe, for instance, not only runs across early hominin culture from *Homo ergaster*, *erectus*, and *heidelbergensis* but also across Africa, Europe, and Asia. An earlier species, *Homo habilis*, had a more crude hand tool, but a tool nonetheless. Marek Kohn and Steven Mithen point out that the hand axe is not simply abundant across the globe but is highly symmetrical and finished, suggesting that beyond any utility the tool functioned to attract mates. The team of historian, archaeologist, and anthropologist Daniel Smail, Mary Stiner, and Timothy Earle agree and say that dazzling artifacts as material goods communicated any range of sexual signals, from trust to prestige. Finely crafted stone tools functioned as a display of the physical energy, manual and visual dexterity, and ingenuity in manufacturing these items. In this way the hand axe, some of which are very large and many of which were never used, served as a reliable fitness indicator and as the aesthetic display of one's intelligence.

The point is that with the production of the improved hand axe by *Homo ergaster* at about 1.8mya combined with his ability to form a symbolic self as per psychologist Constantine Sedikides, these early human beings had *already* developed many significant cognitive aptitudes. These skills, some argue, were related to sexual selection. There was a social function in terms of butchering and sharing meat and learning how to make tools. What we call artistic culture might have origins in human psychology, and not entirely as a fitness mechanism for survival and reproduction, though that factors into any equation.

Artistic culture does not have to be useful but is symbolic with the expressive display of ideas, information, values, and emotions. Art, whether consciously or not, advertises the healthful ingenuity of one's genes or a group's superiority. As has been suggested throughout this book, artistic culture is like a tool in that it operates on the individual, small group, and wider social sphere to influence the thinking and behavior of others. Artistic culture, then, has a *moral* import. When full art and any aesthetic sense began has been debated. Earlier in this book McBrearty and Brooks and then Nowell clearly credit Neanderthals with some artistic culture. *In Human Evolution*, Robin Dunbar is skeptical. Thomas Junker sees true art as arising only after the Neanderthals with cave art by modern human beings.

The neurobiology of what we call art is ancient. In "The Science of Art," neuroscientist V.S. Ramachandran and philosopher William Hirstein say that the deep neurobiological structures of aesthetic experience cross individual and cultural differences. There is a *genetic mechanism* underlying the appreciation of artistic culture. Employing the notion of rasa, from the original Sanskrit, these authors say that an artist, in effect, attempts to reproduce neurally in the object of representation what one would feel and see from the original, even if there is a deliberate distortion. There might be an adaptive reason for creative enterprise, since the biology of vision tends to home in on the essence of something and eliminate distractions or background noise.

Ramachandran and Hirstein go on to say that studies on rats demonstrate how there are likely brain neurons representing certain effects, patterns, shapes, or forms. Artists can activate or "amplify" these neurons (18). In other words, while art might dramatically distort form it simultaneously and effectively captures and accentuates the rasa of whatever is depicted. Moreover, it is well documented, from the 1970s onward, that the primate brain has visual modules devoted to color, depth, and motion. We seem to find satisfaction in being able to recognize patterns and in being able to group "splotches" into an image, say these authors, just as we respond favorably to color and color combinations in flowers (22).

Musicality, too, has an evolutionary basis according to Steven Mithen in "The Music Instinct." Communication happens in various forms, and music is one, as is dance. Music consists of sounds and rhythms, pitches and tones, and in hominins without language, as we currently see in apes and monkeys, vocalizations and gestures provided an array of communication signals. Valérie Dufour et al. demonstrate how a captive chimpanzee exhibited intentional pacing and rhythm in drumming separate from any display context.

Musical sounds and vocalizations, many of which imitate what was heard in nature, are so ingrained in us, that cultures are drawn to and express themselves musically. Mithen suggests that music is part of our biology, with dedicated brain regions evident, and not just cultural. We see, for example, musical expression and enjoyment in babies. Mithen argues bipedalism, which dropped the larynx and thereby increased vocal sounds and breathing, and a decrease in teeth size, which opened the oral cavity for more sounds, fostered music. He says Neanderthals communicated almost exclusively by musical sounds or singing, whereas African *Homo sapiens* split language from music.

Emotions and Aesthetic Experience

The term aesthetic refers to emotions and so why art is often discussed in terms of feeling. Nancy Aiken sees a function of art as evoking strong emotions, from fear to pleasure. In fact, Aiken says that our pupils dilate in fear when exposed to sharp, pointed, angled, or jagged lines, possibly related to an alarm call. What we now call aesthetic is no doubt built upon basic emotional responses. We approve, disapprove, or simply appreciate that which arouses strong responses in us, whether in contemplation or disgust. Visual excitement can be pleasurable. The power of artistic culture over our emotions, which are the guides to our intelligence, is why the arts are moral. Emotions have a capacity to draw us together or pull us apart.

As Ellen Dissanayake says in *Homo Aestheticus*, the aesthetic is not a special after-ingredient sprinkled into us but is, rather, part of human nature. We like to admire what we consider *beautiful* and to seek what we find *meaningful.* Dissanayake makes a very strong case for the innate human tendency to shape something special from what is common. What we now call art is not an extraordinary byproduct but a basic human need to express and control our emotions. The human need for control is evident in our arts, notes Dissanayake, which have, in contrast to nature, a disproportionate amount of geometric shapes and straight lines. Throughout her book, Dissanayake rightly insists that making art is a behavior we require. As I insist, making artistic culture is a moral behavior in the social realm, whether to coordinate a group coherently or for an individual to question and so recalibrate norms, ideas, or the thinking of others.

In spite of the evolution of cooperation as a means of protection from or competition against other groups, many of our adaptations, says Dissanayake, as where and how to live, are not just biological but cultural. We have artistic culture because it is a pleasant sensation, not merely a mentality, and pleasures can bring people together. Dissanayake's contribution is to see art behavior as a way for someone, in attracting others to share or cooperate, to go beyond the ordinary in an act of *making special.* Her reformulation of making special is *artification*, performance with emphasis on aesthetic devices. Artification includes simplification/formalization, repetition, exaggeration, elaboration, and manipulation. This special making of art, though, is not per the Western canon but from an evolutionary perspective regarding cultural activities focused on food resources, life transitions, reproduction, or matters of government. In *The Biological Origins of Art* Nancy Aiken is correct to point out that what we now call art probably had very little to do with ideal beauty. Thomas Scott-Phillips suggests that an artist communicates in a *stylized* display, what's called ostension. Viewers grasp at intent and meaning.

In this regard, let's consider *Feeling Beauty* by G. Gabrielle Starr who examines the neurobiology of aesthetic experience. The book addresses the pressing questions about the relation of emotions to aesthetic experience and how this experience differentiates individuals. Starr makes a convincing analysis to chart neuro-scientifically the very definition, and explore the parameters, of beauty. Fundamental to Starr's argument (as well as to Vessel et al.) is the brain's default mode network, or the brain working on thoughts while at rest from strenuous tasks. There is the self, inward, and others, outward. The default mode network is geared to a process of emotional movement, pleasure and reward, related to aesthetic experience. Key to any aesthetic response is personal relevance.

In large part the *feeling of beauty* means to be moved emotionally on a neural level. Brain matter, even on the synaptic level, is particulate and moves and connects with other synapses. Synaptic connections are created when stimulated. Following aesthetics theorist Elaine Scarry, Starr argues that the "aesthetic value" of literary works and of other arts stems from "images of motion" (8). Images activate neurons albeit differently across individuals. The arts enable one to negotiate, in an attempt at coherence, the onslaught of visual and aural stimuli. Put simply, says Starr, the arts help *shape perception.* As I have been arguing, artistic culture can go beyond directing perception to influencing behavior. There are moral dimensions, therefore, to the arts.

Invoking the eighteenth century philosopher Francis Hutcheson, Starr demonstrates how aesthetics is less about externals and more about personal value judgments. This is what Shaftesbury calls one's feeling of approval or disapproval and which he relates to moral sensations. Starr gets into the how but ignores any adaptive function, as evolutionary biologist Randy Thornhill might. Thornhill would argue for the adaptive function of feeling beauty. Our sensation of beauty is affective, and we respond positively to the look of certain foods, mates, and habitats since in our prehistory our motivation toward the aesthetically pleasing was rewarded. Science philosopher Olaf Breidbach goes further to say that beauty acts like a signal of the "biological reality" of something. We are repulsed by rotting meat for a reason, just as we are attracted to a clear-skinned and symmetrical face.

Emphasizing cultural group selection, Johan De Smedt and Helen De Cruz say we have an art intuition that orients us to artistic concepts. They claim that thinkers like Denis Dutton focus too much on art objects and not on the evolved cognitive mechanisms that help us recognize what we now call art. Hunter-gatherers who do not make art nevertheless recognize drawings depicting hunts. Any creative social depictions carry overtones of the biology of morality.

In terms of creativity, Andrea Kantrowitz says cognition involves perception, movement, and introspection. Moreover, how we perceive and process the world rests on proprioception or the sense of our own body and its movements. Merlin Donald would note how the body self-models representation. Art is a form of introspection we require and enjoy, both for creator and participant. There is nothing static about aesthetic experience. Neurotransmitters can overwhelm a synapse and nearby neurons in highly charged moments of pleasure, intellectual activity, or emotion. Emotionally-charged moments can occur in the creation or viewing of artistic culture.

One of Starr's key points concerns the almost organic, flowing process of how one sees, feels, and contemplates art, what Martha Nussbaum, paraphrased by Starr, calls a *paradigm shift*. For example, the default mode network is implicated in memory, theory of mind, fantasizing, and creativity. Paul Armstrong explores how literary works are a form of pretense and play in the brain, and neuroscientist Semir Zeki has written about how our brains are not averse to embracing, so as to tackle and to accommodate, ambiguity. In other words, Starr claims, an aesthetic experience gives rise to our valuing something or some occurrence over something else. Using the word twice within a span of six pages, Starr says that when the brain encounters and is rewarded by an aesthetic experience, one learns how to qualify likenesses with what at first sight appears "incommensurable" (21, 27).

Although Starr spends time talking about consciousness and cognition, there is no acknowledgment of the adapted mind, and she delineates what she calls the inadequacy of various evolutionary psychologists in addressing individual differences. There is truth in criticizing some of the broad ranging assertions of early writing in evolutionary psychology. While Darwinians often talk on a species level, many aspects of selection operate at the level of the individual. Human brain processes that include cognition, consciousness, and reason are evolved mechanisms advanced from variation as well as competition and inheritance. One of the leading authorities on consciousness, Christof Koch, asserts that the physicality of subjective feelings has provided an evolutionary advantage. Subjectivity is individualistic and adaptive.

In *The Descent of Man* Darwin refers to a *taste for beauty* in the natural world, e.g., bird colors and songs. E.O. Wilson, in his biophilia hypothesis, claims that we too have innate attractions to the natural world. This taste is not just sensory pleasure or simply sexual selection but assumes a degree of judgment, which implies some cognitive faculty. Many male animals send signals, as in bright colors or expansive antlers for attraction or to demonstrate vitality. A few female receivers of the signals decide whom to mate with. The evaluation by the female of the species has an immediate effect on mate choice, and within her offspring are the genes for choice and for the prominent signal. Thus we have in nature the evolution of aesthetics. Richard Prum confirms Darwin's ideas on sexual selection by saying natural selection does not fully account for many features and characteristics in nature. Aesthetic sensibility can function to reinforce brain capacities that impact on our perception of the visual world.

For anthropologist Kathryn Coe, in "Art: The Replicable Unit," art fosters cooperative behaviors through learning and conditioning, and social conduct is important in extending the line of descendants. Going back at least to the Middle Paleolithic period of about 350kya, early people were regularly painting their bodies. The assumption of anthropologists, says Coe, is that human beings respond to colors much as other animals. We share visual mechanisms with mammals, and in terms of mate choice body decoration can serve as a social signal. In discussing the origins of art, Coe says that around 70kya male skulls, in what is now Iraq, were artificially bound, suggesting a costly and painful process merely to attract the attention of a mate.

Other, later behaviors, says Coe, include dental modification, filing, tooth removal, and tooth decoration with metal substances. These practices crossed a span of geography, from Central America, Australia, Asia, Africa, and Europe. This widespread use of body paint and body modification suggests a predisposition for art and any psychological mechanism to attract attention. While female decoration occurs in tribal rites and is associated with mate selection, male decoration, even among their graves, is more pronounced, says Coe. Here is, then, strong support of Darwin's notion of sexual selection regarding aesthetic sensibility. Even today people surgically alter their bodies according to shifting perceptions of beauty.

Nevertheless, Starr says that with a deep view of prehistory evolutionary psychologists are short-sighted in terms of historical cultures and nations. But Darwinists deliberately look at the evolution of culture before the rise of nations. Perhaps this line of thought explains why Starr does not mention, to name only one, Ellen Dissanayake who has written extensively on the origins and prehistory of art. Some authors like Mithen and Klein agree that there is prehistory but then the important neural leap and modular mind circa 50kya that led to the cultural and cognitive flourishing after which Starr and others proceed. Nonetheless, if one looks at McBrearty and Brooks, who question the accuracy of the neural leap, there is much evidence for cultural evolution in, contrary to what Starr suggests, prehistory.

Starr's point is that the arts have the ability to alter human perception and emotion on a cognitive level but not from an evolutionary perspective. In South America, Darwin marveled at what he saw and says, "It creates a feeling of wonder that so much beauty should be apparently created for such little purpose" (qtd. in Browne, *Voyaging*, 216). Remarkably, Darwin at this early time says the beauty is *created*, and only later does he come to realize that the forms, spectacles, sounds, movements, and colors are all a matter of natural and sexual selection. Rather than his then naïve word *purpose*, which hints at teleological design, one might say, rather, function. A humanist will ask, *What is art?* An evolutionist will ask, *Why make art?*

Starr's argument is, however, well taken, for she and others explore and measure subjective aesthetic experiences. Christof Koch asserts that consciousness consists exclusively of physical connections across brain areas and surely is evolutionary. Art is neurobiological, affirming that there is a material, adaptive function for what we call artistic behavior. Relatedly, consciousness is not all it is stacked up to be. Neuroscientist Walter Freeman has described consciousness as a *hurricane*, and Semir Zeki has characterized consciousness as *disunity*. In spite of how well the human mind has evolved, there are disruptive rudiments to the complex networks that give rise to consciousness.

Starr's thrust, broadly speaking, seems to be that while the brain can generate the ability for consciousness, how consciousness manifests itself and changes is determined by the individual experiencing something aesthetically. Aesthetic experience seems to matter to our neurobiology in our willingness to be *absorbed* by art and *abstracted* out of the world, Starr says. Note, though, that in terms of learning, recent studies demonstrate that academic accomplishment flowering from one's entire personality is genetic. For example, see research by Shakeshaft. Starr suggests that brain reward response to some visual, aesthetic stimuli need not be based on evolutionary props of survival and reproduction. Not entirely, considering that we are increasingly learning more about epigenetics. The epigenome is in effect nuclear DNA in the environment that can turn genes on/off. Reward systems, though, are evolutionary, even if mutated, as in our addictions.

Starr indicates that what is visual involves movement. Central to her thesis of the potential of aesthetic experience to energize a revaluation of ideas is the metaphor of motion. Mirror neurons, of course, are implied not only in the motor imagery in arts but also in sympathy. Starr notes how some critics have gone as far as suggesting that when one observes visual or is engaged in literary art there is a sensation of the artist's creative *movements*. Some art that *moves* us, says Starr, triggers mirror neurons and therefore "offers a promising route for modeling...aesthetic pleasures..." (101). Paul Armstrong reports we have so-called canonical neurons which are stimulated not simply by action but by objects that have the potential to act. This means that a property of some motor neurons is in control of our response to artifacts.

Starr takes these notions further by explaining that by its nature beauty "is necessarily about comparison, contrast, integration, and competition..." (117). Indeed, art surprises us and supersedes other thoughts or images. Motion is involved here, too. There can be variant readings of a text even by the same person over time. There can be power-potential or influence in emotional and cognitive moments. Consider how we are constantly re-making artistic culture.

Brain Neurobiology and Art

In "The Neurobiology of Ambiguity," Semir Zeki suggests that our brains like art since there is a challenge to construct and reconstruct images in any number of ways. We invite ambiguity since it helps us interpret and build meaning. The brain as viewer participates in creating a work of art since it searches for regularity and patterns. No doubt thoughtful involvement is an evolutionary adaptation from our hunter and gatherer forebears dealing with a dangerous environment. At the same time the brain might realize that rather than uniformity there is ambiguity and hence more than one meaning. The challenge of addressing and resolving uncertainty helped us build our brains.

Zeki, therefore, proposes that neuro-biologically our brains accept as possible multiple, simultaneous interpretations or, from an adaptive perspective, solutions to a problem. Zeki offers as an example Vermeer's *Girl with a Pearl Earring*, the subject at once coquettish and yet virginal. Here too there is an evolutionary adaptation in terms of reading another's expression or employing theory of mind.

With a slightly different take, Robert Pepperell says that there is a dichotomous nature in representational art. Viewers participate in a conflict of two worlds of awareness. First, there is what is represented in the composition. Second, there is awareness of how the material components are used and come together. Pepperell suggests that the intensity of the aesthetic experience rests in the extent of the dichotomy; what one might expect is compounded by multiple meanings.

In his 1995 Woodhull lecture on the "Visual Brain," Zeki says that form, color, and motion are processed in different parts of the brain. Vision is elemental to our understanding of our world and why one fourth of the brain is devoted to seeing. Visual brain cells are particular in their responsiveness. There is an entire brain area, says Zeki, near the color center, for facial recognition, which helps explain our interest in portraits. Portraiture is a means for artists to capture personality types and universal emotions in particular people. We respond to portraits as we respond to real people in guessing thoughts and emotions. Based on theory of mind, and our evolutionary pro-social traits, we should not be surprised there are so many portrait paintings or paintings and sculptures that include human forms.

Zeki also says in his Woodhull lecture that there is a set of functional aesthetic domains that interact into what we call aesthetics. The brain is bombarded with stimuli, so there is an active, selective process in order to reach a constant. In many respects, says Zeki, the brain constructs the image. A green leaf is so whether in sun, shade, or darkness. Based on prior knowledge the brain colors the leaf whatever the light. So, too, art represents with its patterns and order, even if deliberately off-centered, the brain's search for versions of the essences, meaning, and significance of the world.

On the first viewing of a painting, says Zeki, many brain cells respond, but on subsequent viewings of the same painting there are fewer and fewer active cells. There are cells in certain regions that respond specifically to certain shapes, lines, and movements, or colors on backgrounds. The brain processes color so as to compare objects, and clearly this has an adaptive function relative to food resources. Visual narrowing or homing in makes for a more effective brain.

There are distinct visual systems in the brain, and we are not conscious of the process but only the result. This means that at bottom, as Zeki would say in a number of other papers, there is a disunity of consciousness. Our conscious experience creates order from asynchronous perceptions. Zeki sees art as a variation of the brain working out abstractions. As Charles Stevens says, art abstracts reality according to lines. Just as the brain is an organ to process and evaluate information, there is no accident to how or why it forms abstractions. Memory of specifics is imperfect and abstract, suggests Zeki, providing an outline of essential details. Abstraction is not necessarily a higher brain function per se but one where the particulars of experience are stored and represented as ideas responding to the physical world. Abstraction is a shorthand way for us to gather, store, and recall pertinent information.

Hideaki Kawabta and Zeki, in focusing on the neural correlates of beauty, suggest that in artistic appreciation there is, simply, response to color or motion cues. Then, more complexly, an aesthetic evaluation comes that is highly variable, even at different times in the same individual. Because the sense of beauty is a melding of various inputs judged through experience and mood, the researchers have found that the motor cortex activates in response to images, prompting either retreat or engagement. Movement away from what is perceived as ugly/fearful and toward beautiful/attractive has adaptive value.

Among others we have been considering in this chapter, Zeki's research across the board demonstrates how our artistic culture is not only evolutionary but most likely adaptive. Brain functions can be seen as a natural extension of our primate emotions and cognition.

Theory of Mind

In *Getting Inside Your Head*, cognitive cultural theorist Lisa Zunshine applies the well-accepted concept of theory of mind to popular culture. However, Koch's ideas about consciousness, especially with its physiological and evolutionary aspects, are fundamental to understanding the roots of theory of mind. Zunshine's thesis is that we often manifest an "embodied transparency" (23) in experiencing or estimating someone else's thoughts and feelings, and our greed (her word) to read other people is what has fueled our culture. I'd suggest that which is embodied became, over time, our artistic culture. As Zunshine says, the body exposes feelings.

In fact, the first sentence of Zunshine's book says it all: "We live in other people's heads: avidly, reluctantly, consciously, unawares, mistakenly, inescapably" (xi). Of course *mistaken* since as often as we attempt to read another's mind we are not always correct. Yet since theory of mind is anxiously working, it has enabled our cultural icons. We have novels, theater, movies, television, and visual arts because our own introspection and consciousness have made us eager to peer into other minds and even to share our minds. We want the physicality of another person's thoughts and feelings.

We have an inborn need to read other minds, and that's what cultural representations help satisfy. We expect others to attempt to read our own mind, and so our evolution has endowed our bodies with ways to express mental attributes. Importantly, though, we don't know exactly what the mental state is, and yet we accept as true whatever we can glean or conjecture.

Cultural anthropologist Michael Alvard goes as far as saying that our distinctly complex capacity to imagine others and their motives, needs, and desires as we imagine ourselves is what prompted cultural practices. With theory of mind, however fallible, people were able to offer honest signals to each other and so cooperate. Of course, there were cheaters, and further practices arose to mitigate dishonesty. As psychologists Ara Norenzayan and Will Gervais suggest, in terms of values and beliefs, theory of mind gave rise to a realization that there is an arena separate from but yet in contact with reality, something non-physical, a mentalizing that can be pro-social when shared.

This ability to read mental states works well with fictional characters rather than real people and explains why we have cultural representations. Consciousness in real time is not as effective in connecting all the dots concerning behavior, so we make art to charge our intersubjective thoughts and theory of mind. There are three requirements, especially in fiction, for embodied transparency, says Zunshine. Contrast among characters in a key scene;

transience so that the behavior does not persist continuously; and restraint so that the reader has to guess at the character's inner feelings and thoughts.

For instance, we have the theater, where one goes to read other people as much as or more so than the performance. True, too, of other public events and spectacles where we mind guess. In a museum you see someone contemplating a painting and try to imagine what she is thinking. Movies are particularly well suited for portraying feelings and thoughts, especially those complex characters in key scenes who try to hide emotions, and so we attempt to read into them. Of course with a movie the director can offer close-ups of facial expressions not available in the theater. We process fiction differently than fact, and so there is a moral element if we accept something as true and learn later, to our chagrin, that we have been duped.

Painting is especially complex in terms of theory of mind. For instance, we could be reading the mind of someone in the painting, one absorbed in an act. There could be an image of one contemplating a question, such as a marriage proposal, and we hover and guess. There could be one engaged in a dilemma of some sort not disclosed, and we speculate. The point is that we insert our moral emotions into art deliberately. We want to know as much as we can about other people's lives, issues, and problems, since that is the environment in which we evolved and learned how to survive. If the representations in the painting are too challenging (e.g., surreal art) we consider the mind of the artist. With abstract art we focus on reading our own minds.

Some of the authors working in these areas do not delve deep enough into the evolutionary aspects of cognition. Our brain is clearly an evolved mechanism, evident from comparative work done by primatologists, notably the teams of Cheney and Seyfarth or Parker and Gibson. Working in tandem with evolutionary studies, it is time that we see how being human means we are an evolved species with multiple adaptations to a highly complex social environment, both prehistoric and modern, deeply connected to what we call artistic culture. Our concern about other people's thoughts cuts two ways. We are caring and express empathy. We are also strategizing for ways to minimize costs and increase benefits to ourselves. Hence, artistic culture, one of the highest forms of theory of mind, has moral aspects.

Leda Cosmides and John Tooby say we have evolved mechanisms for problem solving. Specific challenges that would have arisen in our long Pleistocene environment include selecting a mate, acquiring a language, establishing kin ties, and cooperating with others. The question now is how these inborn mechanisms manifest themselves. Of course there is evolutionary precedent for these traits, in the special groupings of, e.g., mountain gorillas. Dian Fossey notes how there are very strong kin ties among mountain gorilla social units where an infant is reared among parents, siblings, and peers. We are not determined by our primate past but it exerts tremendous influence since our particular humanity is premised on our hominid prehistory.

More specifically, our mental processes would ultimately reveal behaviors in artifact production, especially if one considers how artifacts, whether tools or body paints, could be used as signals in sexual selection. We know our cultural mechanisms are evolved because nature would not have been kind to us if we were born with empty minds. Rather, we have innate ways of dealing cooperatively and competitively with the world of other people.

We have genes for a reason, too. Genes structure us according to information that so structured our parents and those who came before us, since they all survived to pass along their genes. Our brains contain a good portion of information-rich genes. Of course there is genetic variation, which helps one better survive and so be able to pass along more fitness enhancing genes. Not everyone needed to be an artist in the Pleistocene, but there were sufficient artistic or material culture tendencies expressed often enough that they were selected for. So any ability for artistic culture has been reproduced over time in the human brain, which explains the human predisposition for making and appreciating artistic culture on various levels.

Zunshine's contribution on theory of mind reveals our adapted abilities and has practical applications for literature and visual arts, drama, and film. Zunshine insists, correctly, that literature is an outgrowth of and consistently lubricates the biological parts of our need to interpret other people. Human creative arts are modes of communication in an environment of human minds. We could qualify minds with the word evolved. In other words, all human minds share a common history, according to the evolutionary psychologists. Art in all its forms is a representation of something natural, alive, and evolved by and for human beings. Art binds the natural and social environments built into our inherited past and stored in our shared human mind.

Science and the Literary Imagination

Not that there has to be a direction, but in the muddied ripples of postmodernist thinking where are studies in the arts and humanities headed? If science is among other things a method, what are the humanities? Is it merely a *byproduct* as Steven Pinker has infamously claimed? As recounted by Stephen Davies, Pinker sees the arts as byproducts of our motivation for pleasure and our technological sense to manufacture pleasure. In the wake of post-structuralism, is there any wonder that the humanities for those outside these disciplines or for those outside of academe appear, dare one say, useless? The tide has turned and the water is clearing. In fact, the debacle resulting from much post-structuralist and postmodern writing has been over for some time. One very promising area of work is human evolutionary studies, especially in terms of the adaptive function of the arts.

Some examples of scientific approaches to the arts and culture include, for example, Patrick Colm Hogan who rightly focuses on universals, via the evolutionists, rather than "cultural and historical specificity" (37) of social constructivists. As a species we are more similar than different. Many cultures have analogies and use metaphors. Certain concepts not only cross cultures but cross species, like depth, height, and space, all of which come into artistic play. In literary terms this means that we find in many literatures universals including allusion, imagery, alliteration, and parallelism.

In *How Literature Plays with the Brain*, literary theorist Paul Armstrong takes a neurobiological account of brain processes that, on the one hand, look for patterns and yet, on the other hand, invite ambiguity. From an evolutionary perspective the ability to manage ambiguity is adaptive since it affords the possibility of establishing new patterns in an ever-changing environment. The human response to the arts is not consistent since the brain itself, to use Armstrong's word, is "decentered" (x). For the brain, engagement with the world and especially with arts and ideas is an important form of play where creative flexibility against rigidity has been the means for the survival and increasing sophistication of the human race. Our neurobiology has evolved so that we *read* our surroundings metaphorically.

Armstrong's thesis is that our neurobiological wiring accepts both *constancy* and *flexibility*. To make his argument he relies on cognitive psychologist Stanislas Dehaene's notion of how the brain *recycles* functions for "object recognition." This includes, as well, Zeki's idea of the neurobiology of ambiguity, Damasio's theory of the as-if body feedback loop, and Giacomo Rizzolatti's research on mirror neurons. With as-if we can influence ourselves into thinking we know something. In ambiguous situations, decisive reactions are simulated from different, previous learning and experience. With mirror neurons another's action can become our own.

According to Armstrong, the scientist must explain differences of interpretive opinions we often find in humanistic discourse, since our species not only enjoys but also values creative works that are ambiguous. But at the same time *explain* is not the most accurate word. Our best and most current fMRI technology cannot render a fully accurate account of what happens in the brain during the process of reading, much less decipher how consciousness derives from brain chemicals and cells. As has been noted but worth repeating, an fMRI might be able to measure hot areas in the brain, but blood flow does not necessarily determine real outcomes. We at least know, says Armstrong, there is no "art neuron" and that aesthetic experience in the presence of visual art or in the process of reading a literary work is spread out across the brain among various functions.

While Jean-Pierre Changeux and Stanislas Dehaene say that *harmony* is the mark of an aesthetic experience, Armstrong asserts that Ramachandran along with Zeki say art can appeal to us with its "distortions" (13). We certainly like regularity but are intrigued by oddity. Part of the answer to this curiosity for the discontinuous, Armstrong tells us, is that we experience art emotionally, not in real life, but *as if* it is real. Artistic culture when non-realistic can be counterfactual, *as if*. That conditional tense is moral. The complex incongruities of art are reflected in the complex mapping response in the human brain. There is no tangible dichotomy between harmony-distortion. Rather, both are parts of a whole neurobiological process of challenge, test, play, and tentative evaluation. We are wired to expect the unexpected.

The brain is a complex organ of multifaceted parts separate and yet connected. As a whole the brain is not a teleological agent, although it prefers organization and patterns. Since the brain evolved over a long period in circumstances different from the past six thousand years or so in which writing developed, our brains have jerry-rigged other functions to help us read. For instance, there is a visual word form area in the brain's visual cortex important for reading. This area is primarily employed in identifying visual arrangements as it is near brain portions implicated in object and facial recognition. More precisely, this visual word form area becomes active when lettering of any kind in any language is introduced. Armstrong suggests that the brain's neurons in this small spot have accommodated themselves to cultural and not only evolutionary forces.

Armstrong spends a good deal of time covering neuronal change through use, disuse, and plasticity. Though controversial, research suggests neurogenesis in some brain regions via history and repetition. Patterns of use/disuse reflect and are reflected in the give-and-take movements of reading. Some neuroscientists like Zeki speak of reward systems in the brain while others might speak of pleasure systems. There is a playful exchange between that which is informational and that which is pleasurable. Evidence of this exchange between information and play can be found in dissonant music and complex, indefinite works of art and novels. Armstrong's point is that this playful interchange is supported by neurobiology itself since the brain's organizational structure is, he reminds us, "decentered" (52). There would be evolutionary advantages in the brain's openness to challenge and stimulation. Information/

play exchange also explains cultural differences in terms of what is presented and accepted as art.

Reading is forward-looking and expectant. One comprehends the whole over time through parts, each of which comes at different times. Does one's interpretation merely exhibit a projection or *reinforcement* of what one believes? More likely reading is a test of beliefs and abilities. Artistic culture can have the effect of altering, even slightly, one's perspective. Can *ambiguity* result in inner conflict so that there is confusion or at best a search for meaning? There are no terminal points in our consciousness because it is always active. Armstrong's idea seems to be that our brain is wired to be tested and so to negotiate many variables. While Zeki says our visual cortex hearkens for a sense of constancy, the human brain is open to varieties of content, and these contents are often unfinished and not tuned finely.

Accepting incongruities permits the brain to admit conflicting data, and hence the notion of play. Recognizing ambiguity and fictional play are adaptive strategies in negotiating the physical and social worlds. We see this tendency to congruity/incongruity in our ability to build metaphors, and that capacity reflects the human brain's willingness to embrace oppositions, to make meaning through creation and destruction. A difficult text or complex image reveals various meanings among individuals by virtue of each brain's history and plasticity. Surely this flexible neurobiology is adaptive, allowing the brain not simply to receive data but also to render decisions about the input.

Armstrong relies on phenomenology to bolster his points. Calling on Heidegger, he says that there is always a space between being in the world and our neurobiological reaction. This gap, however, motivates reflection and generates a crisscross between reader and text. The same could be said for the plastic arts. Play results in back-tracking in order to move forward interpretively. What I've been arguing is that we are capable of participating in two worlds simultaneously. The real world and our feelings. Artistic culture contains and juggles both perspectives. Our analysis and explanation of action, events, and especially human conduct via culture is an expression of our moral emotions. Brain biology is part and parcel of axons, action potentials, cell polarization, refractory periods, synchronous neuronal activity, excitation/relaxation, delta/theta/alpha/beta/gamma waves to demonstrate that what we might ultimately see and feel as meaning is the product of bunched but dissimilar neuronal patterns, says Armstrong. Harmony implies disruption, the latter of which produces action potentials that increase sensitivity to learning. Homeostasis might be the tendency of the body, but stasis is shunned by an active mind.

Armstrong says that what we consider the self is not to be equated with our neurobiology. Instead, our neurobiology is part of the self. Self is not merely a brain state but part of temporal reality, and so his reliance on phenomenology. This thinking discounts, however, cause and effect since an organism can often choose its own best environment. In the case of human beings we've carved out artistic niches that exclude others. Continuing with his theme of de-centeredness, Armstrong goes on to demonstrate or argue, depending on the reader's perspective, that the brain is "a society" of multiple but interlocking "processes" and not necessarily an "individual" (128).

If I have only one brain, no matter how many parts, it is mine. However one defines consciousness, it dies with you. The bottom line is that my neurobiology is my own. On one hand I inherited much of it, on the other I have contributed to making it via thoughts and feelings I've decided to associate with and keep. If there is ultimately no individual, then I am not responsible. Just as we have the notion of the company one keeps impacting on the outcome of conduct, our self-selected feelings and ideas are the company we keep in consciousness. Artistic culture consists of morally challenging and morally satisfying worlds.

Certainly consciousness as even William James knew is messy and continuous. Character as Kant and Schopenhauer knew, in spite of their differences, is multi-dimensional and flexible. Truly, personhood and personality are complex organic forms, and from an evolutionary perspective the somewhat amorphous quality of personality falls in line with variation. Yet Armstrong fails to address who is responsible for the circumstances. The discussion should not simply be about the neurobiology of brain processes but why those processes create different outcomes for different individuals. Why is Van Gogh an artist? Why does so-and-so dislike Van Gogh's work?

This brings Armstrong to theory of mind, simulation theory, and mirror neurons to help us negotiate personal and social emotions. Simulation can occur immediately after birth, whereas theory of mind occurs around age four, since only by then can one understand how others do not share the same beliefs. Critics of simulation theory say the problem is that we are supposedly simulating something we already know. But the upshot is that we surely have simulation and theory of mind capacities, and the *bridge* to both might lie in mirror neurons. There are skeptics who do not place high importance on mirror neurons. But the fact is that discussion of them along with theory of mind and simulation re-centers any debate about the value of artistic culture around the social brain hypothesis and my notion of culture as moral making or challenging the status quo.

Neuroscientist Thierry Chaminade questions statements about so-called mirror neurons, of which, strictly speaking, there are not many. Instead, there is "the concept of 'motor resonance'…" (124). The implications of Chaminade's findings affect accepted beliefs about social learning. A mirror neuron creates a physical response, but the physiological reaction is really via a premotor neuron. Research, says Chaminade, demonstrates that only 6.5% of observed acts and only 5.5% of executed acts are truly the result of mirror neurons. We simply cannot prove "direct correspondence" asserts Chaminade between a mirror neuron and an action, and most of what are called mirror are really premotor neurons reacting to observed action. Chaminade admits that these findings come from research done on captive monkeys working with human caregivers. His point is that neural areas related to observation and the execution of an action are not housed in particular neurons but widely distributed.

Any "equation" of mirror neurons to imitation is questionable, and Chaminade prefers, rather, "motor resonance" where neurons related to motor activity fire up during or in the perception of an action (138). Chaminade's discussion recalls Starr and how arts are not just visual but motor. Indeed, the brain's connecting visual and tactile experiences, in this regard, is not truly a mirror but more a motor effect that enables neuroplasticity and hence learning. Chaminade's idea is that, from an evolutionary standpoint, the brain's capacity for motor skills and not necessarily for abstraction is what led to tool manufacture. This idea means that motor resonance has social cognition implications. One needs *to understand* and not abstractly *mirror* an action.

A point of this book is that there is no one gene for art or music but a suite of genes, brain functions, and environmental pressures that acted on and to which we reacted in our long history to develop artistic culture. The one unifying point, as I see it, is that our social capacity ties everything together, and our social capacity, since it implies right/wrong and good/bad is moral. No one would deem any action by a primate as either moral or immoral. Reciprocal, yes. Altruistic, perhaps. What happens in the natural world is neither good nor bad. But Robert Wright is correct to call *us* a moral animal. By virtue of our executive brain mechanisms, however, our thoughts and deeds can be moral or not, even when directed against another species.

Armstrong suggests that what's key is the notion of alter ego. The paradox of knowing oneself through another, a *doubling capacity*, is clearly part of our psychology and neurobiology. Whether through theory of mind, simulation theory, or mirror neurons, we participate in other minds, with good and bad intentions. These morally tinged mental activities are what account for play engagement in artistic culture. We don't often physically challenge another person, group, or set of values or beliefs. But we can confront others via creating art or participating in artistic culture that asserts our ideas and feelings.

Group Harmony and the Revaluation of Morals

Building off the work of neuroscientist Jean-Pierre Changeux, one could say that variation, competition, and inheritance produced a socially aware brain with compounds of empathy and deception. Because we survived in complex groups that relied on collective knowledge and behaviors, we evolved culture. Nonetheless, the artistic ethos often deliberately rattles one's emotions or challenges a community to realign its values. Subsequently, the jiggling of any group balance reinforces the need for renewed harmony. Homeostasis of the brain/organism is reflected in its general desire for social equanimity.

In spite of variability, there are common denominators concerning social welfare across human cultures, and sociality helped our species survive. Manifestations of care for the survival of the group can be found in artistic culture like ritual, myths, music, and visual arts. Changeux correctly postulates that artistic culture constitutes our easily learned group-specific memories and norms. Pivotal to the evolution of human culture, then, is individual and social consciousness.

How do we quantify a state of consciousness? What makes for artistic consciousness? Consciousness was built on a brain that already had emotions, desires and motives, so in many respects consciousness is a means to access those feelings and thoughts. We are back to the problem of reading neural data on an fMRI and then puzzling over an unexpected outcome in spite of the data. What makes us human is not simply the volume of our brain, since some species have larger brains, but the complex network of and conversation among neurons. From an evolutionary perspective the benefits of a larger brain need to outweigh the energy costs of adapting one. Any benefits came in terms of homeostatic slow development of the brain, especially the neocortex, and hence why large brains are uncommon.

The question about individual consciousness has tremendous impact on our discussion about culture, and specifically about the creation and viewing of artistic culture. Psychologist Tone Roald foregrounds how in an aesthetic experience we need to consider the mind of the individual. More precisely, the subject of an aesthetic encounter is not the work of art but the subjectivity of the viewer. In a similar vein, philosopher Alva Noë de-emphasizes the work of neuroscientists who focus on brain activity in any meeting with art and stresses instead the biography of the individual and the cultural context of any aesthetic response. Noë insists that like philosophy, art is a means for us to question and reorganize our thoughts and lives.

In spite of any species evolved mechanisms and instincts for survival, tension between the complex social brain and individual differences helps account for the continuous surge of artistic culture. An individual creates, sometimes for herself, sometimes for the group. Consciousness can be distinctly individual and yet collective. Culture is a manifestation of how we exhibit, test, and challenge consciousness, whether our own or someone else's. We know we are all different, each with a unique consciousness, and artistic culture capitalizes off these differences. Artistic culture can also capture social consciousness in archetypal forms which give us a positive or negative feeling to accompany the collective sense of a person, place, or event. We need the feeling since it drives a sensation of agency, control. Perhaps we find some mastery in artistic culture.

Since our long evolutionary heritage during the Pleistocene was social, and since a great deal of our consciousness is about other people, our cultural representations mimic what we do in real life. We read characters in novels and guess their intentions, we study the body language of actors on stage for emotional clues and respond to cues from the audience, and we examine closely the facial nuances of screen actors to gauge their thoughts. Beyond any evolutionary or mechanistic function, these social behaviors carry moral import since they impact our way of thinking about others in terms of actions. Being moral is not just following the rules but, as Nietzsche advises, questioning them for revaluation.

My whole point in this book is that human beings, and many primates, are conscious, social, and exhibit intelligence that hinges on what we call *mores*. Even with brain injury, emotions and judgment, to borrow from Antonio Damasio, are *flattened* but not eliminated. One can still know right from wrong.

In spite of any so-called neural leap, which we discussed in Chapter 1, we tend to be a slowly evolving species, as Darwin emphasizes. Thousands of generations were required, on top of multiple adaptations, for us to evolve artistic culture. In large part these adaptive developments are grounded in our kin-centered and social natures. Individuals are concerned, for good and bad motives, about the needs and desires of others. Through mind guessing, mind sharing, and intersubjective psychological communication we evolved arts closely allied to our sociality. So there is a strongly inherent moral aspect in our artistic culture.

Conclusion

In this study we have seen how culture is a manifestation of the mind, and the mind works in both cognitive and emotional ways. That is, like the politics of great apes, most of what we most often do concerns self-interest in terms of others. And so our minds, both conscious and unconscious, are geared toward moral sensations and emotions, those physical and mental feelings that help us interact positively or not with others.

Cognition includes, to some extent, sensation, the interplay between the senses and higher order thinking. Cognition is not necessarily perception but is nonetheless part of the higher thinking process. There is a question, however, especially on an individual basis stemming from temperament and environment, about what remains in memory. Damasio would be the first to say that memories are, because of the complexity of their storage and recollection, mere shadows of what actually happened. What does one choose to do, to see, attend to? In part, this is individualized character, but there is a large impact in how one's cultural environment can draw and shape our attention. The groups we are in can determine some of our cultural constructs, and hence why we have notions about *the company one keeps*.

Cultural cognition can work on two planes. First, there are concepts derived from representational phenomena held prototypically in mind. Then there can be generalized patterns or schemas derived from personal experience and one's contact with culture. Conceptual knowledge, both innate and learned, is essential for understanding the external world. But problem solving comes in terms of behavior. What we now call social cognition, in this discussion having included moral emotions, makes sense from an evolutionary perspective. First, there was the self in relation to other selves. Then, as groups grew and splintered to become other groups, the brain needed a way to assess and store the character traits of others. Eventually, groups literally and figuratively marked themselves in certain ways, and these markings and behavior led to cultural manifestations we now call art.

The aesthetic response to art is universal, but the expressions of artistic creation are culturally different. Impressions are compared with previous or similar impressions in memory and then an evaluation is made. There is a combination of cultural representations and universal, human mental processes, or how general knowledge is represented and stored in the brain. Representations from memory are patterns, suggesting that data stored in different parts of the brain can be scrambled to establish new patterns. The pattern only signifies what all the small pieces mean as a whole. Because our

 | DOI 10.1163/9789004319486_007

brain evolution has established both conceptual and pattern mechanisms, groups can significantly modify or change symbols culturally in a short time.

As this book has demonstrated, for the most part because of a common ancestor and shared evolution, there are human universals. Nearly anyone can enter another culture and participate in cultural, commercial practices, and learn the language. However, as the argument of this book has also demonstrated, on a finer scale, there might be strong resistance to particular values and beliefs by the fact that there are many splinters from the generalized human culture. The same cognitive and emotional underpinnings of being human can be expressed culturally in dramatically different ways.

There are processes of cognition outside of awareness, like type-casting or biases, which affect behavior. Partly, this bias is evolutionary in how we are equipped to make automatic judgments about many particular habits and movements of an individual or group. However, the individual character can, based on values and beliefs, override any automatic trait inferences. Motivation, then, is affected by cognition. Emotions cause us to alter cognitive processes and therefore have a dramatic effect on representations and memory. Since many cognitive mechanisms are evolutionary, and since emotional expressions are also evolutionary, the foundations of culture are in our evolutionary past. In a broad sense, we are cultural creatures who share and yet challenge values, beliefs, and practices. We place ourselves in a social context, we perceive ourselves so placed holding a history of self, which in turn then affects some social cognition and behavior.

In order to be social, each of us needs a concept of self. We perceive ourselves as we perceive others and even objects, with social and non-social concepts. How we perceive ourselves and others is infused with moral emotions and cognitive mechanisms, which, by virtue of our group living, has given rise to cultural creation and communication.

Bibliography

Adams, Jill U. 2008. "Obesity, Epigenetics, and Gene Regulation." *Nature Education*. 1.1. 128. Web.

Aiken, Nancy. 1989. "Evolution of the Human Psyche." *The Human Revolution*. Chris Stringer and Paul Mellars, eds. Edinburgh: U Edinburgh pp. 455–513.

———. 1990. *How Did Humans Evolve? Reflections on the Uniquely Unique Species*. Ann Arbor, MI: U Michigan.

———. 1998. *The Biological Origins of Art*. Westport, CT: Praeger.

———. 2013."Aesthetics and Evolution." *Aisthesis. Pratiche, linguaggi esaperi dell'estetico* 6.2. 61–73. Web.

Alexander, Richard. 1987. *The Biology of Moral Systems*. NY: Aldine de Gruyter.

Allchin, Douglas. 2009. "The Evolution of Morality." *Evo Edu Outreach* 2. 590–601. Web.

Alvard, Michael S. 2003. "The Adaptive Nature of Culture." *Evolutionary Anthropology* 12. 136–149.

Antón, Susan C., Richard Potts, and Leslie C. Aiello. 2014. "Evolution of Early *Homo*: An Integrated Biological Perspective." *Science* 345.6192. 1236828–1–1236828–13. Web.

Aron, Elaine N. and Arthur Aron. 1997. "Sensory-Processing Sensitivity and Its Relation to Introversion and Emotionality." *Journal of Personality and Social Psychology* 73.2. 345–368. Print.

Asao, Kelly and David M. Buss. 2016. "The Tripartite Theory of Machiavellian Morality: Judgment, Influence, and Conscience as Distinct Moral Adaptations." *The Evolution of Morality*. Todd K. Shackelford and Ranald D. Hansen, eds. Springer. 3–25.

Atran, Scott. 2010. "The Evolution of Religion: How Cognitive By-Products, Adaptive Learning Heuristics, Ritual Displays, and Group Competition Generate Deep Commitments to Prosocial Religions." *Biological Theory* 5.1. 18–30.

Balter, Michael. 2009. "On the Origin of Art and Symbolism." *Science* 323. 709–711. Web.

Barkow, Jerome, et al., eds. 1992. *The Adapted Mind: Evolutionary Psychology and the Generation of Culture*. NY: Oxford UP.

Barnard, Alan. 1999. "Modern Hunter-Gatherers and Early Symbolic Culture." *The Evolution of Culture*. Robin Dunbar, Chris Knight, Camilla Powers, eds. New Brunswick: Rutgers UP.

Batson, C. Daniel. 2014. "Are the 'Principles in His Nature, Which Interest Him in the Future of Others,' Moral?" Putnam, Hilary, Susan Neiman, Jeffrey P. Schloss, eds. *Understanding Moral Sentiments: Darwinian Perspectives*. New Brunswick, NJ: Transaction Publishers. 43–57.

Baumeister, Roy F. 2005. *The Cultural Animal: Human Nature, Meaning, and Social Life*. Oxford: OUP.

Bechara, Antoine. 2004. "The Role of Emotion in Decision-Making: Evidence from neurobiological patients with orbitofrontal damage." *Brain and Cognition* 55. 30–40.

Bednarik, Robert G. 2013. "Pleistocene Palaeoart Art of Asia." *Arts* 2. 46–76. Web.

Berger, Lee R. et al. 2015. "*Homo Naledi*, a New Species of the Genus Homo from the Dinaledi Chamber, South Africa." *eLife* 4:e09560. Web.

Berman, Sophie. 2013. "Science Sans Conscience N'est Que Ruine de l'Âme." *ASEBL Journal* 9.1. 36–39. Web.

Bloom, Paul. 2004. *Descartes' Baby: How the Science of Child Development Explains What Makes Us Human*. NY: Basic Books.

Boehm, Christopher. 2012. *Moral Origins: The Evolution of Virtue, Altruism, and Shame*. NY: Basic Books.

———. 2014. "Defining the Evolutionary Conscience: Finishing a Job that Darwin Started." Putnam, Hilary, Susan Neiman, Jeffrey P. Schloss, eds. *Understanding Moral Sentiments: Darwinian Perspectives*. New Brunswick, NJ: Transaction Publishers. 27–41.

Bouchard, Thomas J., Jr. 1994. "Genes, Environment, and Personality." *Science* 264. 1700–1701.

——— and John C. Loehlin. 2001. "Genes, Evolution, and Personality." *Behavior Genetics* 31.3. 243–273.

———. 2004. "Genetic Influence on Human Psychological Traits." *Current Directions in Psychological Science*. 13.4. 148–151.

Boyd, Robert, Peter J. Richerson, Joseph Henrich. 2011. "The Cultural Niche: Why Social Learning is Essential for Human Adaptation." *Proceedings of the National Academy of Sciences* 108.2. 10918–10925. Web.

Breibach, Olaf. 2003. "The Beauties and the Beautiful – Some Considerations from the Perspective of Neuronal Aesthetics." *Evolutionary Aesthetics*. Eckart Voland and Karl Grammer, eds. Berlin: Springer. 39–68.

Browne, Janet. 1995. *Charles Darwin: Voyaging*. Princeton: PUP.

Buskes, Chris. 2013. "Darwinism Extended: A Survey of How the Idea of Cultural Evolution Evolved." *Philosophia* 41. 661–691. Web.

Buss, David M. 2008. *Evolutionary Psychology: The New Science of the Mind*. Third edition. Boston: Pearson.

Carruthers, Peter. 2013. "The Distinctively-Human Mind: The Many Pillars of Cumulative Culture." *Evolution of Mind, Brain, and Culture*. Gary Hatfield and Holly Pittman, eds. Philadelphia: U of Pennsylvania Museum of Archaeology and Anthropology. 325–345.

Chalmers, David J. 2003. "Consciousness and its Place in Nature." *Blackwell Guide to Philosophy of Mind*. S. Stich and T. Warfield, eds. Web.

Chaminade, Thierry. 2013. "Functions of Premotor Cortices: From Motor Control to Social Cognition." *Evolution of Mind, Brain, and Culture*. Gary Hatfield and

Holly Pittman, eds. Philadelphia: U of Pennsylvania Museum of Archaeology and Anthropology. 123–148.

Changeux, Jean-Pierre. 2002. *The Physiology of Truth: Neuroscience and Human Knowledge*. Cambridge, MA: Belknap/Harvard UP.

Chase, Philip G. 2013. "Human Culture Is More Than Memes and Transmission." *Evolution of Mind, Brain, and Culture*. Gary Hatfield and Holly Pittman, eds. Philadelphia: U of Pennsylvania Museum of Archaeology and Anthropology. 347–378.

Chauvet, Jean-Marie, Eliette Brunel Deschamps, Christian Hillaire. 1996. *Dawn of Art: The Chauvet Cave*. NY: Harry N. Abrams.

Cheney, Dorothy L. and Robert M. Seyfarth. 1990. *How Monkeys See the World: Inside the Mind of Another Species*. Chicago: U Chicago P.

Churchland, Patricia S. 2011. *Braintrust: What Neuroscience Tells Us About Morality*. Princeton: PUP.

Cieri, Robert L., et al. 2014. "Craniofacial Feminization, Social Tolerance, and the Origins of Behavioral Modernity." *Current Anthropology* 55.4. 419–443.

Clay, Zanna et al. 2015. "Functional Flexibility in Wild Bonobo Vocal Behavior." *PeerJ* 3.e1124. Web.

Clottes, Jean. 2013. "Why Did They Draw in Those Caves?" *Time and Mind: The Journal of Archaeology, Consciousness and Culture* 6.1. 7–14. Web

Coe, Kathryn. 1992. "Art: The Replicable Unit – An Inquiry Into the Possible Origin of Art as a Social Behavior." *Journal of Social and Evolutionary Systems* 15.2. 217–234.

———. 2003. *The Ancestress Hypothesis: Visual Art as Adaptation*. New Brunswick, NJ: Rutgers UP.

Conkey, Margaret W. 2001. "Hunting for Images, Gathering up Meanings: Art for Life in Hunting-Gathering Societies." *Hunter-Gatherers: An Interdisciplinary Perspective*. Catherine Painter-Brick, Robert H. Layton, Peter Rowley-Conwy, eds. Cambridge: CUP. 267–291.

Consoli, Gianluca. 2015. "Early Art and the Evolution of Grounded Emotions." *Aisthesis. Pratiche, linguaggi esaperi dell'estetico* 8.1. 147–156. Web.

Copp, David. 2008. "Darwinian Skepticism About Moral Realism." *Philosophical Issues* 18. 186–206.

Cosmides, Leda and John Tooby. 1989. "Evolutionary Psychology and the Generation of Culture." Part two. *Ethology and Sociobiology* 10. 51–97.

———. 1992. "Cognitive Adaptations for Social Exchange." *The Adapted Mind: Evolutionary Psychology and the Generation of Culture*. Jerome Barkow, et al., eds. NY: Oxford UP. 163–228.

———. 1997. "Evolutionary Psychology: A Primer." www.psych.ucsb.edu/research/cep/primer.html Univ. of California, Santa Barbara, Center for Evolutionary Psychology. Web.

Coss, Richard G. 2003. "The Role of Evolved Perceptual Biases in Art and Design." *Evolutionary Aesthetics*. Eckart Voland and Karl Grammer, eds. Berlin: Springer. 69–130.

Coward, Fiona and Clive Gamble. 2008. "Big Brains, Small Worlds: Material Culture and the Evolution of Mind." *Philosophical Transactions of the Royal Society B* 363. 1969–1979. Web.

Coyne, Jerry A. 2009. *Why Evolution is True*. NY: Penguin.

Crane, Mary Thomas. 2010. *Introduction to Cognitive Cultural Studies*. Lisa Zunshine, ed. Baltimore: Johns Hopkins UP. 103–114.

Damasio, Antonio. 1994. 2005. *Descartes' Error: Emotion, Reason, and the Human Brain*. NY: Penguin.

———. 1999. *The Feeling of What Happens: Body and Emotion in the Making of Consciousness*. Orlando: Harcourt/Harvest.

———. 2003. *Looking for Spinoza*. Orlando: Harcourt.

———. 2012. *Self Comes to Mind: Constructing the Conscious Brain*. NY: Vintage.

Darwin, Charles. 2003 (1859). *On the Origin of Species*. Joe Carroll, ed. Ontario, CN: Broadview.

———. 1871. 1879, second ed. 2004. *The Descent of Man*. London: Penguin Books.

Davies, Stephen. 2012. *The Artful Species: Aesthetics, Art, and Evolution*. Oxford: OUP.

Dawkins, Richard. 2006. *The Selfish Gene*. 1976. Oxford: OUP.

Deacon, Terrence W. 1997. *The Symbolic Species: The Co-Evolution of Language and the Brain*. NY: Norton.

———. 2006. "The Aesthetic Faculty." *The Artful Mind: Cognitive Science and the Riddle of Human Creativity*. Oxford: OUP. 21–53.

De Smedt, Johan, and Helen De Cruz. 2012. "Human Artistic Behaviour: Adaptation, Byproduct, or Cultural Group Selection?" *Philosophy of Behavioural Biology*. Kathryn S. Plaisance and Thomas Reydon, eds. Springer. 167–187.

de Waal, Frans. 1996. *Good Natured: The Origin of Right and Wrong in Humans and Other Animals*. Cambridge, MA: Harvard UP.

———. 1999. "Anthropomorphism and Anthropodenial: Consistency in our Thinking about Humans and Other Animals." *Philosophical Topics* 27.1. 255–280.

———. 2006. With Robert Wright, Christine M. Korsgaard, Philip Kitcher, and Peter Singer. *Primates and Philosophers: How Morality Evolved*. Stephen Macedo, ed. Princeton: PUP.

———. 2007 (1982). *Chimpanzee Politics: Power and Sex Among Apes*. Baltimore, MD: Johns Hopkins UP.

———. 2009. *The Age of Empathy: Nature's Lessons For a Kinder Society*. NY: Harmony Books.

———. 2013. *The Bonobo and the Atheist: In Search of Humanism Among the Primates*. NY: Norton.

Dirks, Paul H.G.M. et al. 2015. "Geological and Taphonomic Context for the New Hominin Species *Homo naledi* from the Dinaledi Chamber, South Africa." *eLife* 4:e09561. Web.

Dissanayake, Ellen. 1995. *Homo Aestheticus: Where Art Comes From and Why*. Seattle: U Washington P.

———. 1998. "Komar and Melamid Discover Pleistocene Taste." *Philosophy and Literature* 22.2. 486–496. Web.

———. 2001. "Aesthetica Incunabula." *Philosophy and Literature* 25. 335–346.

———. 2001. "Becoming *Homo Aestheticus*: Sources of Aesthetic Imagination in Mother-Infant Interactions." *SubStance* 94/95. 85–103.

———. 2015. "Aesthetic Devices Promote Viewers' Felt Emotional Connections with Artists." *ASEBL Journal* 11.2. 37–39. Web.

Donald, Merlin. 1993. "Précis of *Origins of the Modern Mind: Three Stages in the Evolution of Culture and Cognition*." *Behavioral and Brain Sciences* 16.4.737-791.

———. 2013. "Mimesis Theory Re-Examined, Twenty Years after the Fact." *Evolution of Mind, Brain, and Culture*. Gary Hatfield and Holly Pittman, eds. Philadelphia: U of Pennsylvania Museum of Archaeology and Anthropology. 169–192.

Douglas, Pamela Heidi and Liza R. Muscovice. 2015. "Pointing and Pantomime in Wild Apes? Female Bonobos Use Referential and Iconic Gestures to Request Genito-genital Rubbing." *Scientific Reports* 5.13999. Web.

Dufour, Valérie, et al. 2015. "Chimpanzee Drumming: A Spontaneous Performance with Characteristics of Human Musical Drumming." *Scientific Reports* 5.11320. Web.

Dunbar, Robin 1998. "The Social Brain Hypothesis." *Evolutionary Anthropology* 6.5. 178–190.

———. 2004. "Gossip in Evolutionary Perspective." *Review of General Psychology* 8.2. 100–110.

———. 2014. *Human Evolution: A Pelican Introduction*. London: Penguin.

Dunbar, Robin, Christ Knight and Camilla Power, eds. 1999. *The Evolution of Culture*. New Brunswick: Rutgers UP.

Dutton, Denis. 2009. *The Art Instinct: Beauty, Pleasure, and Human Evolution*. NY: Bloomsbury.

Easterlin, Nancy. 2010. "Cognitive Ecocriticism: Human Wayfinding, Sociality, and Literary Interpretation." *Introduction to Cognitive Cultural Studies*. Baltimore, MD: The Johns Hopkins UP. 257–274.

Flack, Jessica C. and Frans B.M. de Waal. 2002. "'Any Animal Whatever': Darwinian Building Blocks of Morality in Monkeys and Apes." *Evolutionary Origins of Morality: Cross-Disciplinary Perspectives*. Leonard D. Katz, ed. Thorverton, UK: Imprint Academic. 1–29.

Flinn, Mark V., David C. Geary, Carol V. Ward. 2005. "Ecological Dominance, Social Competition, and Coalitionary Arms Races: Why Humans Evolved Extraordinary Intelligence." *Evolution and Human Behavior* 26. 10–46.

Fossey, Dian. 1983. 2000. *Gorillas in the Mist*. Boston: Houghton Mifflin.

Freeman, Walter J. 2006. "Consciousness, Intentionality, and Causality." *Does Consciousness Cause Behavior?* Susan Pockett, William P. Banks, and Shaun Gallagher, eds. Cambridge, MA: MIT pp. 73–105.

Gabora, Liane and A. Russon. 2011. "The Evolution of Human Intelligence." *The Cambridge Handbook of Intelligence*. R. Sternberg and S. Kaufman, eds. Cambridge: CUP. 328–350.

Galdikas, Biruté and Nancy Briggs. 1999. *Orangutan Odyssey*. NY: Harry N. Abrams.

Gärdenfors, Peter. 2013. "The Role of Cooperation in the Evolution of Protolanguage and Language." *Evolution of Mind, Brain, and Culture*. Gary Hatfield and Holly Pittman, eds. Philadelphia: U of Pennsylvania Museum of Archaeology and Anthropology. 193–216.

Gardner, Howard. 2011. *Frames of Mind: The Theory of Multiple Intelligences*. Third edition. NY: Basic Books.

Gazzaniga, Michael. 2005. *The Ethical Brain*. NY: Dana Press.

Gibson, Kathleen Rita. 1990. "New Perspectives on Instincts and Intelligence: Brain Size and the Emergence of Hierarchical Mental Construction Skills." *"Language" and Intelligence in Monkeys and Apes: Comparative Developmental Perspectives*. Sue Taylor Parker and Kathleen Rita Gibson, eds. Cambridge: Cambridge UP. 97–128.

Gómez, Juan Carlos. 1990. "The Emergence of Intentional Communication as a Problem-solving Strategy in the Gorilla." *"Language" and Intelligence in Monkeys and Apes: Comparative Developmental Perspectives*. Cambridge: Cambridge UP. 333–355.

Gómez-Robles, Aida, et al. 2015. "No Known Hominin Species Matches the Expected Dental Morphology of the Last Common Ancestor of Neanderthals and Modern Humans." *PNAS* 110.45. 18196–18201. Web.

Gonzales, Lauren, A., et al. 2015. "Cerebral Complexity Preceded Enlarged Brain Size and Reduced Olfactory Bulbs in Old World Monkeys." *Nature Communications* 6.7580. Web.

Goodall, Jane. 1986. *The Chimpanzees of Gombe: Patterns of Behavior*. Cambridge: Harvard UP.

Gottschall, Jonathan and David Sloan Wilson, eds. 2005. *The Literary Animal: Evolution and the Nature of Narrative*. Evanston, IL: Northwestern UP.

Greenfield, Patricia Marks and E. Sue Savage-Rumbaugh. 1990. "Grammatical Combination in *Pan paniscus*: Processes of Learning and Invention in the Evolution and Development of Language." *"Language" and Intelligence in Monkeys and Apes: Comparative Developmental Perspectives*. Sue Taylor Parker and Kathleen Rita Gibson, eds. Cambridge: Cambridge UP. 540–578.

Greene, Joshua. 2013. *Moral Tribes: Emotion, Reason, and the Gap Between Us and Them*. NY: Penguin.

Haidt, Jonathan. 2001. "The Emotional Dog and Its Rational Tail: A Social Intuitionist Approach to Moral Judgment." *Psychological Review* 108.4. 814–834.

———. 2003. "The Moral Emotions." *Handbook of Affective Sciences*. R.J. Davidson, K.R. Scherer, and H.H. Goldsmith, eds. Oxford: OUP. 852–870.

———. 2006. *The Happiness Hypothesis: Finding Modern Truth in Ancient Wisdom*. NY: Basic Books.

Haig, David. 1995. "Prenatal Power Plays." *Natural History* 104.12. 39.

———. 1997. "The Social Gene." *Behavioral Ecology: an Evolutionary Approach*. J.R. Krebs and N.B. Davies, eds. London: Blackwell. 284–304.

———. 2003. "On Intrapersonal Reciprocity." *Evolution and Human Behavior* 24. 418–425. Web.

Hamilton, W.D. 1964. "The Genetic Evolution of Social Behavior II." *Journal of Theoretical Biology* 7. 17–52.

Hardy, Charlie and Mark Van Vugt. 2006. "Nice Guys Finish First: The Competitive Altruism Hypothesis." *Personality and Social Psychology Bulletin* 32. 1402–1413.

Harmand, Sonia et al. 2015. "3.3-Million-year-old Stone Tools from Lomekwi 3, West Turkana, Kenya." *Nature* 521. 31–315. Web.

Hatfield, Gary and Holly Pittman, eds. 2013. *Evolution of Mind, Brain, and Culture*. Philadelphia: U of Pennsylvania Museum of Archaeology and Anthropology.

Hauser, Marc. 2000. *Wild Minds: What Animals Really Think*. NY: Henry Holt.

———. 2006. *Moral Minds: How Nature Designed Our Universal Sense of Right and Wrong*. NY: Harper Collins.

Heinrich, Joseph and Natalie Heinrich. 2006. "Culture, Evolution and the Puzzle of Human Cooperation." *Cognitive Systems Research*. 1–26. Web.

Henshilwood, Christopher S., et al. 2011. "A 100,000-Year-Old Ochre-Processing Workshop at Blombos Cave, South Africa." *Science* 334. 219–222. Web.

Herman, David. 2010. "Narrative Theory after the Second Cognitive Revolution." *Introduction to Cognitive Cultural Studies*. Lisa Zunshine, ed. Baltimore: Johns Hopkins UP. 155–175.

Hey, Jody. 2013. "What Genetics Can Tell Us about the Origins of the Modern Human Brain." *Evolution of Mind, Brain, and Culture*. Gary Hatfield and Holly Pittman, eds. Philadelphia: U of Pennsylvania Museum of Archaeology and Anthropology. 91–104.

Heyes, Cecilia. 2003. "Four Routes of Cognitive Evolution." *Psychological Review* 110.4. 713–727.

Hogan, Patrick Colm. 2010. "Literary Universals." *Introduction to Cognitive Cultural Studies*. Lisa Zunshine, ed. Baltimore: Johns Hopkins UP. 37–60.

———. 2010. "On Being Moved: Cognition and Emotion in Literature and Film." *Introduction to Cognitive Cultural Studies*. Baltimore, MD: The Johns Hopkins UP. 237–256.

Humphrey, Nicholas. 1983. *Consciousness Regained: Chapters in the Development of Mind*. Oxford: OUP.

———. 1988. "The Social Function of Intellect." 1976. *Machiavellian Intelligences*. Richard Byrne and Andrew Whiten, eds. Oxford: Clarendon pp. 303–317.

Ishizu, Tomohiro and Semir Zeki. 2011. "Toward a Brain-Based Theory of Beauty." *Plos One* 6.7. E21852. 1–10. Web.

Jablonka, Eva and Marion J. Lamb. 2005. *Evolution in Four Dimensions: Genetic, Epigenetic, Behavioral, and Symbolic Variation in the History of Life*. Cambridge, MA: MIT/ Bradford.

Jackson, Timothy P. 2014. "Thirty-five Years Among the Genes: A Twainian Take on Moral Anthropology and Reductive Biology." Putnam, Hilary, Susan Neiman, Jeffrey P. Schloss, eds. *Understanding Moral Sentiments: Darwinian Perspectives*. New Brunswick, NJ: Transaction Publishers. 169–201.

Janson, H.W. 1991. *History of Art*. Fourth edition, revised by Anthony F. Janson. NY: Harry N. Abrams.

Junker, Thomas. 2010. "Art as a Biological Adaptation, or: Why Modern Humans Replaced the Neanderthals." *Quartär* 57. 171–178.

Kagan, Jerome and Nancy Snidman. 2004. *The Long Shadow of Temperament*. Cambridge, MA: Belknap P of Harvard UP.

———. 2006. *An Argument for Mind*. New Haven: Yale UP.

———. 2007. *What Is Emotion? History, Measures, and Meanings*. New Haven: Yale UP.

———. 2009. *The Three Cultures: Natural Sciences, Social Sciences, and the Humanities in the 21st Century*. Cambridge: CUP.

———. 2010. *The Temperamental Thread: How Genes, Culture, Time, and Luck Make Us Who We Are*. NY: Dana P.

Kantrowitz, Andrea. 2012. "The Man Behind the Curtain: What Cognitive Science Reveals about Drawing." *The Journal of Aesthetic Education* 46.1. 1–14.

Kaplan, Hillard, Kim Hill, Jane Lancaster, A. Magdalena Hurtado. 2000. "A Theory of Human Life History Evolution: Diet, Intelligence, and Longevity." *Evolutionary Anthropology* 9. 156–185.

Kappeler, Peter M., Joan Silk, Judith Burkart, Carel P. van Schaik. 2010. "Primate Behavior and Human Universals: Exploring the Gap." *Mind the Gap: Tracing the Origins of Human Universals*. Peter M. Kappeler and Joan B. Silk, eds. Heidelberg: Springer.

Kawabta, Hideaki and Semir Zeki. 2004. "Neural Correlates of Beauty." *Journal of Neurophysiology* 91. 1699–1705.

Klein, Richard. 2009. *The Human Career*. Third edition. Chicago: U Chicago P.

Knight, Chris, Camilla Powers, Ian Watts. 1995. "The Human Symbolic Revolution: A Darwinian Account." *Cambridge Archaeological Journal* 5.1. 75–114.

Koch, Christof. 2012. *Consciousness: Confessions of a Romantic Reductionist*. Cambridge, MA: MIT Press.

Köhler, Wolfgang. 1973 (1925). *The Mentality of Apes*. Ella Winter, trans. London: Routledge and Kegan Paul.

Kohn, Marek and Steven Mithen. 1999. "Handaxes: Products of Sexual Selection?" *Antiquity* 73. 518–526. Web.

Koops, Kathelijne, Elisabett Visalberghi, Carel P. van Schaik. 2014. "The Ecology of Primate Material Culture." *Biology Letters* 10. Web.

——— et al. 2015. "Chimpanzees and Bonobos Differ in Intrinsic Motivation for Tool Use." *Scientific Reports* 5.11356. Web.

——— et al. 2015. "Cultural Differences in Ant-Dipping Tool Length Between Neighbouring Chimpanzee Communities at Kalinzu, Uganda." *Scientific Reports* 5.12456. Web.

Krebs, Dennis L. 2011. *The Origins of Morality: An Evolutionary Account*. Oxford: OUP.

Kuhn, Steven L. and Mary C. Stiner. "Middle Paleolithic 'Creativity.'" *Creativity in Human Evolution and Prehistory*. Steven Mithen, ed. London: Routledge, 1998. 143–164.

———, Mary Stiner, David S. Reese, and Erksin Gülec. 2001. "Ornaments of the Earliest Upper Paleolithic: New Insights from the Levant." *PNAS* 98.13. 7641–7646. Web.

Lahti, David C. and Bret S. Weinstein. 2005. "The Better Angels of Our Nature: Group Stability and the Evolution of Moral Tension." *Evolution and Human Behavior* 26. 47–63.

———. 2014. "On the Partnership Between Natural and Moral Philosophy." Putnam, Hilary, Susan Neiman, Jeffrey P. Schloss, eds. *Understanding Moral Sentiments: Darwinian Perspectives*. New Brunswick, NJ: Transaction Publishers. 229–256.

Latzman, Robert D. et al. 2015. "Neuroanatomical Correlates of Personality in Chimpanzees (*Pan troglodytes*): Associations Between Personality and Frontal Cortex." *NeuroImage* 123. 63–71. Web.

LeDoux, Joseph. 1996. *The Emotional Brain: The Mysterious Underpinnings of Emotional Life*. NY: Simon and Schuster.

Lee, Richard B. and Irven DeVore. 1968. *Man the Hunter*. Chicago: Aldine.

Lobo, Ingrid. 2008. "Environmental Influences on Gene Expression." *Nature Education*. 1.1. 39. Web.

Lock, Anthony. 2015. "Evolutionary Aesthetics, the Interrelationship Between Viewer and Artist, and New Zealandism." *ASEBL Journal* 11.2. 3–36.

Marcus, Gary. 2009. "How *Does* the Mind Work? Insights from Biology." *Topics in Cognitive Science* 1: 145–172.

Marshall-Pescini, Sarah, Andrew Whiten. 2008. "Chimpanzees (*Pan troglodytes*) and the Question of Cumulative Culture: An Experimental Approach." *Animal Cognition* 11. 449–456.

Mayer, Emeran A. 2011. "Gut Feelings: The Emerging Biology of Gut-Brain Communication." *Nature Reviews* 12. 453–466. Web.

McBrearty, Sally and Alison S. Brooks. 2000. "The Revolution that Wasn't: A New Interpretation of the Origin of Modern Human Behavior." *Journal of Human Evolution* 39. 453–563.

McConachie, Bruce. 2010. "Toward a Cognitive Cultural Hegemony." *Introduction to Cognitive Cultural Studies*. Lisa Zunshine, ed. Baltimore: Johns Hopkins UP. 134–150.

McGrew, William C. 1998. "Culture in Nonhuman Primates?" *Annual Review of Anthropology*. 27. 301–328.

———. 2010. "Chimpanzee Technology." *Science* 328. 579–580.

Mesoudi, Alex. 2011. *Cultural Evolution: How Darwinian Theory Can Explain Human Culture and Synthesize the Social Sciences*. Chicago: U Chicago P.

Meyer, Matthias, et al. 2015. *Proceedings of the 5th Annual Meeting of the European Society for the Study of Human Evolution*. London. 162.

Miles, H. Lyn White. 1990. "The Cognitive Foundations for Reference in a Signing Orangutan." *"Language" and Intelligence in Monkeys and Apes: Comparative Developmental Perspectives*. Sue Taylor Parker and Kathleen Rita Gibson, eds. Cambridge: Cambridge UP. 511–539.

Miller, Geoffrey. 2001. *The Mating Mind: How Sexual Choice Shaped the Evolution of Human Nature*. NY: Anchor.

Miller, Jon D., Eugenie C. Scott, Shinji Okamoto. 2006. "Public Acceptance of Evolution." *Science* 313. 765–766. Web.

Mithen, Steven. 1996. *The Prehistory of the Mind: The Cognitive Origins of Art, Religion and Science*. NY: Thames and Hudson.

———. 2009. "The Music Instinct: The Evolutionary Basis of Musicality." *The Neurosciences and Music III – Disorders and Plasticity: Annals of the N.Y. Academy of Sciences* 1169. 3–12.

———. 2013. "The Cathedral Model for the Evolution of Human Cognition." *Evolution of Mind, Brain, and Culture*. Gary Hatfield and Holly Pittman, eds. Philadelphia: U of Pennsylvania Museum of Archaeology and Anthropology. 217–233.

Moulton, Samuel T. and Stephen M. Kosslyn. 2009. "Imagining Predictions: Mental Imagery as Mental Emulation." *Philosophical Transactions of the Royal Society B* 364. 1273–1280. Web.

Neiman, Susan. 2014. "Setting Aside Facts." Putnam, Hilary, Susan Neiman, Jeffrey P. Schloss, eds. *Understanding Moral Sentiments: Darwinian Perspectives*. New Brunswick, NJ: Transaction Publishers. 213–228.

Noë, Alva. 2015. *Strange Tools: Art and Human Nature*. NY: Hill and Wang.

Norenzayan, Ara and Will M. Gervais. 2011. "The Cultural Evolution of Religion." *Creating Consilience: Integrating the Sciences and the Humanities*. Edward Slingerland and Mark Collard, eds. Oxford: OUP. 243–265.

Nowell, April. 2013. "Cognition, Behavioral Modernity, and the Archaeological Record of the Middle and Early Upper Paleolithic." *Evolution of Mind, Brain, and Culture*. Gary Hatfield and Holly Pittman, eds. Philadelphia: U of Pennsylvania Museum of Archaeology and Anthropology. 235–262.

Nussbaum, Martha C. 2014. "Compassion: Human and Animal." Putnam, Hilary, Susan Neiman, Jeffrey P. Schloss, eds. *Understanding Moral Sentiments: Darwinian Perspectives*. New Brunswick, NJ: Transaction Publishers. 123–150.

Pagel, Mark. 2012. *Wired for Culture: Origins of the Human Social Mind*. NY: W.W. Norton.

Palmer, Alan. 2010. "Storyworlds and Groups." *Introduction to Cognitive Cultural Studies*. Lisa Zunshine, ed. Baltimore: Johns Hopkins UP. 176–192.

Parker, Sue Taylor. 1990. "Why Big Brains are so Rare: Energy Costs of Intelligence and Brain Size in Anthropoid Primates." *"Language" and Intelligence in Monkeys and Apes: Comparative Developmental Perspectives*. Sue Taylor Parker and Kathleen Rita Gibson, eds. Cambridge: Cambridge UP. 129–154.

Parker, Sue Taylor and Kathleen Rita Gibson, eds. 1990. *"Language" and Intelligence in Monkeys and Apes: Comparative Developmental Perspectives*. Cambridge: Cambridge UP.

Penn, Derek C., Keith J. Holyoak, Daniel J. Povinelli. 2008. "Darwin's Mistake: Explaining the Discontinuity Between Human and Nonhuman Minds." *Behavioral and Brain Sciences* 31. 109–178. Web.

Pepperell, Robert. 2015. "Artworks as Dichotomous Objects: Implications for the Scientific Study of Aesthetic Experience." *Frontiers in Human Neuroscience* 9:295. Web.

Pinker, Steven. 2012. *The Better Angels of Our Nature: Why Violence Has Declined*. NY: Penguin.

———. 2014. "The Moral Instinct." Putnam, Hilary, Susan Neiman, Jeffrey P. Schloss, eds. *Understanding Moral Sentiments: Darwinian Perspectives*. New Brunswick, NJ: Transaction Publishers. 59–80.

Polderman, Tinca J.C., et al. 2015. "Meta-analysis of the Heritability of Human Traits Based on Fifty Years of Twin Studies." *Nature Genetics*. Web.

Pope, Stephen J. 2014. "Moral Decision Making: Interpreters of Darwin and Thomas Aquinas." Putnam, Hilary, Susan Neiman, Jeffrey P. Schloss, eds. *Understanding Moral Sentiments: Darwinian Perspectives*. New Brunswick, NJ: Transaction Publishers. 151–168.

Povinelli, Daniel J. and Jennifer Vonk. 2003. "Chimpanzee Minds: Suspiciously Human?" *Trends in Cognitive Sciences* 7.4. 157–160.

Preston, Stephanie and Frans B.M. de Waal. 2002. "Empathy: Its Ultimate and Proximate Bases." *Behavioral and Brain Sciences* 25. 1–71.

Preuschoft, Signe. 2000. "Primate Faces and Facial Expressions." *Social Research* 67.1. 245–271.

Prum, Richard O. 2012. "Aesthetic Evolution by Mate Choice: Darwin's *Really* Dangerous Idea." *Philosophical Transactions of the Royal Society B* 367. 2253–2265. Web.

———. 2013. "Coevolutionary Aesthetics in Human and Biotic Artworlds." *Biology and Philosophy*. Web.

Putnam, Hilary. 2014. "Not Very Much." Putnam, Hilary, Susan Neiman, Jeffrey P. Schloss, eds. *Understanding Moral Sentiments: Darwinian Perspectives*. New Brunswick, NJ: Transaction Publishers. 203–211.

Putnam, Hilary, Susan Neiman, Jeffrey P. Schloss, eds. 2014. *Understanding Moral Sentiments: Darwinian Perspectives*. New Brunswick, NJ: Transaction Publishers.

Radovčić, Davorka, et al. 2015. "Evidence for Neanderthal Jewelry: Modified White-Tailed Eagle Claws at Krapina." *Plos One* 10.3. Web.

Ramachandran, V.S. and William Hirstein. 1999. "The Science of Art: A Neurological Theory of Aesthetic Experience." *Journal of Consciousness Studies* 6.6/7. 15–51.

Richardson, Alan. 2010. "Facial Expression Theory from Romanticism to the Present." *Introduction to Cognitive Cultural Studies*. Lisa Zunshine, ed. Baltimore: Johns Hopkins UP. 65–83.

Richerson, Peter J. and Robert Boyd. 2005. *Not By Genes Alone: How Culture Transformed Human Evolution*. Chicago: U Chicago P.

———. 2013. "Rethinking Paleoanthropology: A World Queerer Than We Supposed." *Evolution of Mind, Brain, and Culture*. Gary Hatfield and Holly Pittman, eds. Philadelphia: U of Pennsylvania Museum of Archaeology and Anthropology. 263–302.

Ridley, Matt. 1996. *The Origins of Virtue: Human Instincts and the Evolution of Cooperation*. NY: Penguin Books.

Rimfeld, Kaili, Yulia Kovas, Philip S. Dale, and Robert Plomin. 2015. "Pleiotrophy Across Academic Subjects at the End of Compulsory Education." *Scientific Reports* 5.11713. Web.

Roald, Tone. 2015. *The Subject of Aesthetics: A Psychology of Art and Experience*. Leiden: Brill.

Roberts, Gilbert. 1998. "Competitive Altruism: From Responsibility to the Handicap Principle." *Proceedings of the Royal Society of London: Biological Sciences* 265.1394. 427–431.

Rule, Nicholas O. 2014. "Cultural Neuroscience: A Historical Introduction and Overview." *Online Readings in Psychology and Culture* 9.2. Web.

Schloss, Jeffrey P. 2014. "Darwinian Explanations of Morality: Accounting for the Normal but not the Normative." Putnam, Hilary, Susan Neiman, Jeffrey P. Schloss, eds. *Understanding Moral Sentiments: Darwinian Perspectives*. New Brunswick, NJ: Transaction Publishers. 81–121.

Schurr, Theodore G. 2013. "When Did We Become Human? Evolutionary Perspectives on the Emergence of the Modern Human Mind, Brain, and Culture." *Evolution of Mind, Brain, and Culture*. Gary Hatfield and Holly Pittman, eds. Philadelphia: U of Pennsylvania Museum of Archaeology and Anthropology. 45–89.

Scott-Phillips, Thomas C. 2015. "What is Art? A Pragmatic Perspective" *Think 40*, 14. 87–91. Web.

Sedikides, Constantine and John J. Skowronski. 1997. "The Symbolic Self in Evolutionary Context." *Personality and Social Psychology Review* 1.1. 80–102. Web.

Sedikides, Constantine, John J. Skowronski, and R.I.M. Dunbar. 2006. "When and Why Did the Human Self Evolve?" *Evolution and Social Psychology*. Mark Schaller, Jeffry A. Simpson, and Douglas T. Kenrick, eds. NY: Psychology Press. 25 pgs. Web.

Seeley, W.P. 2013. "Art, Meaning, and Perception: A Question of Methods for a Cognitive Neuroscience of Art." *British Journal of Aesthetics* 53.4. 443–460.

Seyfarth, Robert M. and Dorothy L. Cheney. 2013. "The Primate Mind before Tools, Language, and Culture." *Evolution of Mind, Brain, and Culture*. Gary Hatfield and Holly Pittman, eds. Philadelphia: U Pennsylvania pp. 105–121.

Shakeshaft, Nicholas G. et al. 2013. "Strong Genetic Influence on a UK Nationwide Test of Educational Achievement at the End of Compulsory Education at Age 16." *Plos One* 8.12. 1–10. E80341. Web.

Shryock, Andrew and Daniel Lord Smail, eds. 2011. *Deep History: The Architecture of Past and Present*. Berkeley: U California P.

Shryock, Andrew, Thomas R. Trautmann, Clive Gamble. 2011. "Imagining the Human in Deep Time." Shryock, Andrew and Daniel Lord Smail, eds. *Deep History: The Architecture of Past and Present*. Berkeley: U California pp. 21–52.

Silk, Joan. 2014. "The Roots of Prosocial Preferences." Putnam, Hilary, Susan Neiman, Jeffrey P. Schloss, eds. *Understanding Moral Sentiments: Darwinian Perspectives*. New Brunswick, NJ: Transaction Publishers. 9–26.

Skarsaune, Knut Olav. 2011. "Darwin and Moral Realism: Survival of the Iffiest." *Philosophical Studies* 152. 229–243.

Smail, Daniel Lord and Andrew Shryock. 2011. "Body." Shryock, Andrew and Daniel Lord Smail, eds. *Deep History: The Architecture of Past and Present*. Berkeley: U California pp. 55–77.

Smail, Daniel Lord, Mary C. Stiner, Timothy Earle. 2011. "Goods." Shryock, Andrew and Daniel Lord Smail, eds. *Deep History: The Architecture of Past and Present*. Berkeley: U California pp. 219–241.

Snowdon, Charles T., and Toni E. Ziegler. 2015. "Variation in Prolactin is Related to Variation in Sexual Behavior and Contact Affiliation." *Plos One* 10.3. E0120650. 1–11. Web.

Sober, Elliott and David Sloan Wilson. 1998. *Unto Others: The Evolution and Psychology of Unselfish Behavior*. Cambridge, MA: Harvard UP.

Soriano, Sylvian et al. 2015. "The Still Bay and Howiesons Poort at Sibudu and Blombos: Understanding Middle Stone Age Technologies." *Plos One* 10.7. Web.

Spolsky, Ellen. 2010. "Making 'Quite Anew': Brain Modularity and Creativity." *Introduction to Cognitive Cultural Studies*. Lisa Zunshine, ed. Baltimore: Johns Hopkins UP. 84–102.

Starr, G. Gabrielle. 2010. "Multisensory Imagery." *Introduction to Cognitive Cultural Studies*. Baltimore, MD: The Johns Hopkins UP. 275–291.

———. 2013. *Feeling Beauty: The Neuroscience of Aesthetic Experience*. Cambridge, MA: MIT Press.

Steen, Francis. "A Cognitive Account of Aesthetics." 2006. *The Artful Mind: Cognitive Science and the Riddle of Human Creativity*. Mark Turner, ed. Oxford: OUP. 57–71.

Sterelny, Kim. 2013. "Human Behavioral Ecology, Optimality, and Human Action." *Evolution of Mind, Brain, and Culture*. Gary Hatfield and Holly Pittman, eds. Philadelphia: U of Pennsylvania Museum of Archaeology and Anthropology. 303–324.

Stevens, Charles F. 2001. "Line Versus Color: The Brain and the Language of Visual Arts." *The Origins of Creativity*. Karl. Pfenninger and Valerie R. Shubik, eds. Oxford: OUP. 177–189.

Street, Sharon. 2006. "A Darwinian Dilemma for Realist Theories of Value." *Philosophical Studies* 127. 109–166.

Swirski, Peter. 2013. *From Literature to Biterature: Lem, Turing, Darwin, and Explorations in Computer Literature, Philosophy of Mind, and Cultural Evolution*. Montreal: McGill-Queen's UP.

Tague, Gregory F. 2001. "The Peculiar Morality of the Artist." *Consciousness, Literature and the Arts*. 2.2. Web.

———. 2014. *Making Mind: Moral Sense and Consciousness in Philosophy, Science, and Literature*. Amsterdam: Brill/Rodopi.

Texier, Jean-Pierre et al. 2010. "A Howiesons Poort Tradition of Engraving Ostrich Eggshell Containers Dated to 60,000 Years Ago at Diepkloof Rock Shelter, South Africa." *PNAS* 107.14. 6180–6185. Web.

Thornhill, Randy. 2003. "Darwinian Aesthetics Informs Traditional Aesthetics." *Evolutionary Aesthetics*. Eckart Voland and Karl Grammer, eds. Berlin: Springer. 9–35.

Tomasello, Michael. 1990. "Cultural Transmission in the Tool Use and Communicatory Signalling of Chimpanzees?" *"Language" and Intelligence in Monkeys and Apes: Comparative Developmental Perspectives*. Sue Taylor Parker and Kathleen Rita Gibson, eds. Cambridge: Cambridge UP. 274–311.

———. 2014. *A Natural History of Human Thinking*. Cambridge, MA: Harvard UP.

Tooby, John and Leda Cosmides. 1989. "Evolutionary Psychology and the Generation of Culture." Part one. *Ethology and Sociobiology* 10. 29–49.

———. 1992. "The Psychological Foundations of Culture." *The Adapted Mind: Evolutionary Psychology and the Generation of Culture*. Jerome H. Barkow et al., eds. NY: Oxford UP. 19–136.

———. 2001. "Does Beauty Build Adapted Minds? Toward an Evolutionary Theory of Aesthetics, Fiction and the Arts." *SubStance* 94/95: 6–27.

———. 2005. "Conceptual Foundations of Evolutionary Psychology." *The Handbook of Evolutionary Psychology*. David M. Buss, ed. Hoboken, NY: Wiley. 5–67.

Tooby, John and Irven DeVore. 1987. "The Reconstruction of Hominid Behavioral Evolution Through Strategic Modeling." *The Evolution of Human Behavior: Primate Models*. Warren G. Kinzey, ed. Albany: SUNY P. 183–237.

Trivers, Robert. 1971. "The Evolution of Reciprocal Altruism." *The Quarterly Review of Biology* 46.1. 35–57.

———. 1972. "Parental Investment and Sexual Selection." *Sexual Selection and the Descent of Man*. Bernard Campbell, ed. Chicago: Aldine. 136–179.

Turner, Mark, ed. 2006. *The Artful Mind: Cognitive Science and the Riddle of Human Creativity*. Oxford: OUP.

Turner, Robert and Charles Whitehead. 2008. "How Collective Representations Can Change the Structure of the Brain." *Journal of Consciousness Studies* 15.10/11. 43–57. Web.

van Hooff, Jan A.R.A.M. and Bas Lukkenaar. 2015. "Captive Chimpanzee Takes Down a Drone: Tool Use Toward a Flying Object." *Primates*. Web.

van Leeuwen, Edwin J.C. et al. 2014. "A Group Specific Arbitrary Tradition in Chimpanzees (*Pan troglodyges*)" *Animal Cognition* 17. 1421–1425.

van Schaik, Carel P., et al. 2003. "Orangutan Cultures and the Evolution of Material Culture." *Science* 299. 102–105. Web.

Van Vugt, Mark and Paul A.M. Van Lange. 2006. "The Altruism Puzzle." *Evolution and Social Psychology*. Mark Schaller, Jeffry A. Simpson, and Douglas T. Kenrick, eds. NY: Psychology P. 237–261. Web.

Vermeule, Blakey. 2010. "Machiavellian Narratives." *Introduction to Cognitive Cultural Studies*. Baltimore, MD: The Johns Hopkins UP. 214–230.

Vessell, Edward A., G. Gabrielle Starr and Nava Rubin. 2012. "The Brain on Art: Intense Aesthetic Experience Activates the Default Mode Network." *Frontiers in Human Neuroscience* 6.66. 1–17. Web.

Villmoare, Brian et al. 2015. "Early *Homo* at 2.8 Ma from Ledi-Geraru, Afar, Ethiopia." *Sciencexpress* 10.1126. Web.

Visalberghi, Elisabetta and Dorothy Munkenbeck Fragaszy. 1990. "Do Monkeys Ape?" *"Language" and Intelligence in Monkeys and Apes: Comparative Developmental Perspectives*. Sue Taylor Parker and Kathleen Rita Gibson, eds. Cambridge: Cambridge UP. 247–273.

Voland, Eckart and Karl Grammer, eds. 2003. *Evolutionary Aesthetics*. Berlin: Springer.

von Petzinger, Genevieve, April Nowell. 2014. "A Place in Time: Situating Chauvet within the Long Chronology of Symbolic Behavioral Development." *Journal of Human Evolution* 74. 37–54. Web.

von Rohr, Claudia Rudolf, et al. 2015. "Chimpanzees' Bystander Reactions to Infanticide: An Evolutionary Precursor of Social Norms?" *Human Nature* 26. 143–160. Web.

Warneken, Felix. 2013. "The Origins of Human Cooperation from a Developmental and Comparative Perspective." *Evolution of Mind, Brain, and Culture*. Gary Hatfield and

Holly Pittman, eds. Philadelphia: U of Pennsylvania Museum of Archaeology and Anthropology. 149–168.

West-Eberhard, Mary Jane. 1987. "Flexible Strategy and Social Evolution." *Animal Societies: Theories and Facts*. Y. Ito, J.L. Brown, and J. Kikhawa, eds. Japan Science Society, Tokyo. 35–51

———. 1989. "Phenotypic Plasticity and the Origins of Diversity." *Annual Review of Ecological Systems*. 20. 249–278.

Whiten, Andrew, et al. 2009. "Emulation, Imitation, Over-imitation and the Scope of Culture for Child and Chimpanzee." *Philosophical Transactions of the Royal Society B*. 364. 2417–2428. Web.

———. 2010. "Ape Behavior and the Origins of Human Culture." *Mind the Gap: Tracing the Origins of Human Universals*. Peter M. Kappeler and Joan B. Silk, eds. Heidelberg: Springer. 429–450.

Wilson, D.S. 1983. "The Group Selection Controversy: History and Current Status." *Annual Review of Ecological Systems* 14.159–187.

Wright, Robert. 1994. *The Moral Animal: Evolutionary Psychology and Everyday Life*. NY: Pantheon.

Yerkes, Robert M. 1916. *The Mental Life of Monkeys and Apes: A Study of Ideational Behavior*. Boston: Henry Holt.

Zeki, Semir. 1997. "Visual Art and the Visual Brain." The Woodhull Lecture 1995. *Proceedings of the Royal Institution of Great Britain* 68. 29–63. Web.

———. 2001. "Artistic Creativity and the Brain." *Science* 293.5527. 51–52. Web.

———. 2002. "Neural Concept Formation and Art: Dante, Michelangelo, Wagner." *Journal of Consciousness Studies* 9.3. 53–76.

———. 2003. "The Disunity of Consciousness." *Trends in Cognitive Science* 7.5. 214–218. Web.

———. "The Neurobiology of Ambiguity." 2006. *The Artful Mind: Cognitive Science and the Riddle of Human Creativity*. Oxford: OUP. 243–270.

Zunshine, Lisa, ed. 2010. *Introduction to Cognitive Cultural Studies*. Baltimore, MD: The Johns Hopkins UP.

———2012. *Getting Inside Your Head: What Cognitive Science Can Tell Us About Popular Culture*. Baltimore, MD: The Johns Hopkins University Press.

Index

www.ingramcontent.com/pod-product-compliance
Lightning Source LLC
Chambersburg PA
CBHW031326230926
51848CB00001B/106
9789004305366